ORGANIC SPECTROSCOPY

ORGANIC SPECTROSCOPY

D. W. Brown, A. J. Floyd and M. Sainsbury
School of Chemistry,
University of Bath

JOHN WILEY & SONS
Chichester · New York · Brisbane · Toronto · Singapore

Copyright © 1988 by John Wiley & Sons Ltd.

All rights reserved.

No part of this book may be reproduced by any means, or transmitted, or translated into a machine language without the written permission of the publisher.

Library of Congress Cataloging-in-Publication Data:

Brown, D. W. (David W.), 1937–
   Organic spectroscopy/D. W. Brown, A. J. Floyd, and M. Sainsbury.
   p.   cm.
   Includes bibliographies and index.
   ISBN 0-471-91911-X   ISBN 0-471-91912-8 (pbk.)
   1. Spectrum analysis. 2. Chemistry, Organic. I. Floyd, A. J.
II. Sainsbury, M. III. Title.
QD272.S6B76 1988
547.3′0858—dc 19

British Library Cataloguing in Publication Data:

Brown, D. W.
   Organic spectroscopy.
   1. Organic compounds. Spectroscopy
   I. Title II. Floyd, A. J. III. Sainsbury, M.
   547.3′0858

ISBN 0 471 91911 X (cloth)
ISBN 0 471 91912 8 (paper)

Phototypesetting by Thomson Press (India) Limited, New Delhi
Printed by Bath Press Ltd., Bath, Avon

CONTENTS

Preface		vii
Chapter 1	INTRODUCTION	1
Chapter 2	ULTRAVIOLET AND VISIBLE SPECTROSCOPY	3
	2.1 Basic Principles and Laws	3
	2.2 Presentation of Data	5
	2.3 Instrumentation and Cell Design	5
	2.4 Solvents	6
	2.5 Chromophores, Auxochromes and the Effects of Conjugation	6
	2.6 Selected Chromophoric Systems	8
	Bibliography	23
Chapter 3	INFRARED SPECTROSCOPY	24
	3.1 Measurement of Spectra	24
	3.2 Absorption of Infrared Energy	25
	3.3 Practical Uses of Infrared Spectroscopy	28
	3.4 Problem Solving	30
	3.5 The Infrared Spectra of Organic Compounds	30
	Bibliography	53
Chapter 4	NUCLEAR MAGNETIC RESONANCE (NMR) SPECTROSCOPY	54
	4.1 Introduction	54
	4.2 The NMR Parameters	54
	4.3 Chemical Shift	56
	4.4 Multiplicity; Spin–Spin Splitting; $n+1$ Rule	59
	4.5 Pascal's Triangle	59
	4.6 Coupling Constants	60
	4.7 Commonly Encountered Coupled Systems	60
	4.8 Interpretation of Proton NMR Spectra	79
	4.9 Basic Concepts	85

4.10	Instrumentation	90
4.11	^{13}C NMR Spectroscopy	93
4.12	Simplification of Spectra	107
4.13	Effect of Chirality on the NMR Spectrum	112
4.14	Dynamic Effects Observed by NMR	115
4.15	Problems	121
4.16	Appendices	124
	Bibliography	134

Chapter 5 MASS SPECTROMETRY   135

5.1	Introduction	135
5.2	Production of Spectra	135
5.3	Determination of Molecular Formulae	140
5.4	Fragmentation Processes	147
5.5	Fragmentations Associated with Functional Groups	152
5.6	Interpretation of a Mass Spectrum	175
5.7	Examples of the Interpretation of Mass Spectra	179
	Bibliography	182

Chapter 6 PROBLEMS   184

6.1	Notes on Problems	184
6.2	Problems	186
6.3	Answers to Problems	245

Index   247

PREFACE

In the last decade or so there have been major advances in the instrumental techniques available to the organic chemist. Fourier transform methods applied to both infrared and nuclear magnetic resonance spectroscopy have reduced the time taken to obtain spectra and also reduced the amount of sample required. Larger magnets also allow much greater resolution of complex NMR spectra, and this is further aided by two-dimensional presentations of the data. New NMR spectrometers embody computerization which enables related spin–spin systems to be identified graphically, and to indicate those nuclei which are spatially close.

Similar advances have emerged in mass spectrometry. Chemical ionization and fast atom bombardment techniques provide us with the ability to identify the molecular ions of macromolecules and also those of relatively unstable molecules which under electron impact simply undergo fragmentation.

Already extremely complex problems of structure determination are routinely undertaken by chemists and, since we now need to know much more than just, say, the rigid structures of enzymes and similar biologically important molecules, there is no doubt that further developments in spectroscopic instrumentation and techniques can be expected. In this book, we aim to provide students with a good training in the basics of organic spectroscopy so that they will be able to cope with more complicated and esoteric problems in the future. The whole approach is based on the experience we have gained over the years in teaching undergraduates at the University of Bath.

A considerable proportion of the text is given over to a series of problems, some more difficult than others, which should serve to sharpen the skills we seek to impart.

Finally we wish to express our particular gratitude to Springer-Verlag and to the authors of *Spectral Data for Structure Determination of Organic Compounds*, Springer-Verlag (1983), for their permission to base much of the data in Tables 10, 11, 12, 14, 19, 20 and 21 upon those presented in the above book. We are also indebted to Varian Associates and the Aldrich Chemical Co. Ltd., for their permission to reproduce NMR spectra from their catalogues and to Chris Cryer, Harry Hartell and David Wood of the University of Bath for their help in recording many of the mass and NMR spectra used in this book.

CHAPTER 1

Introduction

Whenever an organic compound is synthesized or encountered for the first time, there is a need to determine its structure. To achieve this, the compound is subjected to spectral analysis, most often employing the four techniques described in this book: electronic, infrared and nuclear magnetic resonance spectroscopy and mass spectrometry. The results obtained from this combined analysis are normally sufficient to allow at least the gross structure of an 'unknown' to be deduced and often its relative stereochemistry to be worked out.

The approach can be summarized as follows: (a) the molecular formula can be obtained from the mass spectrum, and the fragmentation pattern noted; (b) the nature of the functional groups is available from the infrared spectrum; (c) whether these functional groups are in conjugation or not is ascertained from the electronic spectrum; and (d) possible structural alternatives can be selected and/or discounted by considering the environments of the hydrogen and carbon nuclei present in the molecule from the data given in the ^{1}H and ^{13}C NMR spectra. Finally, the detailed information from each method is re-examined to ensure that a self-consistent solution to the problem is obtained.

In practice, of course, the nature of the chemical reactions leading up to the 'unknown' are often understood, which helps to reduce the number of structural possibilities. Also, it is frequently the case that the problem is resolved by the judicious application of just one, or two, of the techniques, with perhaps the aid of a simple chemical test and a melting or boiling point determination. In more complex examples, however, all the information is required and here the chemist is continually cross-checking the conclusions drawn from one set of data with those from another. Indeed, such a critical approach is the key to success, and one of the high points of the working day is when all the individual pieces of the jigsaw suddenly come together to provide an unambiguous answer to a difficult structural problem.

Throughout the book we have concentrated on the utilization of the spectral data, without much consideration of the instrumentation required to obtain them. This 'black-box' attitude has its drawbacks and we hope that the interested student will read the textbooks cited in the various bibliographical sections to make up for this deficiency. From a practical point of view it is essential to take note of the conditions which have been used to determine the spectra. For example, it is necessary to observe whether solvent peaks need to be subtracted from the spectrum, or whether the chart has been deliberately off-set to accommodate signals exhibited outside the normal instrumental range.

Throughout the book numerous examples are cited to illustrate how the information from individual techniques can be used to the maximum effect, and at the end we have provided a set of problems to test how well the reader can integrate this knowledge to solve 'unknown' structures.

CHAPTER 2

Ultraviolet and Visible Spectroscopy

2.1 BASIC PRINCIPLES AND LAWS

2.1.1 Absorption of electromagnetic radiation

Photons from the ultraviolet and visible regions of the electromagnetic spectrum have sufficient energy to promote electrons from their ground states in organic molecules to excited states. The difference in energy between any two of these states is quantized so that only a photon of precisely the right energy may be absorbed.

The energy content (E) of the photon is related to the wavelength of the incident radiation and is given by the expression

$$E = hc/\lambda$$

where h (Planck's constant) = 6.63×10^{-34} joule seconds (J s); c (the velocity of light) = 3×10^8 metre seconds^{-1} (m s^{-1}) and λ is the wavelength (in metres).

Clearly, photons from the ultraviolet region have higher energies than those from the visible region, and indeed exposure to ultraviolet radiation of short wavelength (below 190 nm) can be sufficient to bring about the excitation of an electron from a single (σ) bond. In this event bond fragmentation occurs, but generally spectroscopists are less interested in this type of effect (which is better studied by mass spectrometry) than in observing either the promotion of an electron from the pi (π) system of a multiple bond, or from a non-bonded (n) pair into an associated antibonding orbital.

Such processes do not involve the break-up of the molecule and occur at wavelengths in the range 190–750 nm. This includes the 'accessible' ultraviolet and the visible regions of the electromagnetic spectrum. Should measurements be required at shorter wavelengths then specialized instrumentation is needed (see Section 2.3).

Promotion of a π-electron into a corresponding antibonding π^* orbital is a commonly observed event when unsaturated molecules are irradiated and, although a non-bonded (n or lone pair) electron, by definition, does not have a corresponding excited state, relatively little energy is involved in its promotion

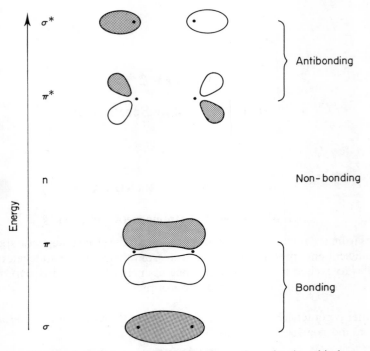

Figure 2.1 Relative energies of σ, π, π^* and σ^* orbitals (schematic).

into a π^* orbital (see the next section). Not surprisingly, therefore, the spectra of many compounds show bands which correspond to n–π^* transitions. In addition, certain n–σ^* promotions are possible, but the energies involved are much higher and these absorptions are only observed at wavelengths below 190 nm.

The relative energies of the various bonding, non-bonding and antibonding orbitals are shown schematically in Fig. 2.1.

2.1.2 Probability factors and the Beer–Lambert relationship

The question of whether a particular absorption band is observed or not does not depend simply on the energy content of the incident light. Thus, although an electronic transition may be energetically favourable the associated band is sometimes not observed, or if present in the spectrum it is of very low intensity. The n–π^* transition of carbonyl compounds represents a good example: here the energy gap between the orbitals bearing the lone pair electrons on the oxygen atom and π^* is relatively small but, since the lone pair orbital is orientated orthogonally to the excited state orbital, the overlap between them is minimal.

Good orbital overlap is needed if an electron is to move from one orbital to another, so it is not surprising that the intensity of the n–π* bands of carbonyl compounds are very weak. Such bands are often described as 'forbidden transitions' and the fact that they are observed at all is a consequence of the uncertainty of locating an electron in space and the effect of molecular vibrations which may, in some cases, distort the molecular structure and hence the relative orientations of the orbitals. Selection rules related to molecular symmetry also apply, as they do in infrared spectroscopy (see p. 26).

Ultraviolet and visible spectra are easily measured on solutions, and the Beer–Lambert law is then used to relate the amount of electromagnetic radiation absorbed to the concentration of the sample dissolved in a suitable solvent and contained in a cell of known dimensions. This law is expressed in the form

$$\log I^0/I = \varepsilon c l$$

where I^0 is the intensity of the incident beam, I is the intensity of the transmitted beam, after passage through a cell of length l cm, and c is the concentration of the sample in moles per litre (mol l^{-1} or M). The molar absorptivity, ε (mol l^{-1} cm^{-1}), is thus a probability factor associated with the 'allowedness' or 'forbiddenness' of the particular transition being observed.

2.2 PRESENTATION OF DATA

The usual form of an electronic spectrum is a plot of transmittance ($\log I^0/I$) against wavelength in nm (10^{-9} m), and from this the ε values can then be calculated if the molar concentration is known. Often ε, or log ε, is then manually replotted against wavelength. The logarithmic plot is most useful when bands of varying intensity occur together in the same spectrum, for whereas 'allowed' transitions may have ε values measured in tens of thousands, 'forbidden' transitions may have values of less than 100.

Electronic absorption spectra may appear as a series of smooth bands rather than as sharp peaks, since each electronic transition is associated with certain rotational and vibrational changes which are taking place at the same time. In non-polar solvents the fine structure of the bands is often visible, but in polar media solvent–solute interactions obscure these details and then a more or less continuous curve is obtained.

Bands are identified by their absorption maxima (λ_{max}), and sometimes when the bands are overlapping the minima are also quoted. If the overlap is severe then only shoulders in the curves may be visible, and here it is customary to record the wavelength at the points of maximum inflection.

2.3 INSTRUMENTATION AND CELL DESIGN

The working range for most spectrometers is from 190 to 750 nm. The light from

190 to 450 nm is provided by a deuterium discharge lamp and the remaining wavelengths by a tungsten source. The light is split into two beams, one of which passes through the cell containing the sample in solution, and the other through a matched cell containing the solvent only. Transmitted light from both beams is then compared automatically and the data are plotted as described in the previous section. Cells constructed from glass are transparent to visible but not to ultraviolet light. Consequently quartz cells, transparent across the whole range, are employed. The normal operational lower limit of 190 nm is set, since below this wavelength air and solvents increasingly absorb ultraviolet radiation.

Measurements in the far-UV region require specialized equipment and are usually conducted on samples in the vapour phase.

2.4 SOLVENTS

Numerous solvents may be used in measuring electronic absorption provided, of course, that they are transparent to light over the desired range of wavelengths. The most commonly employed medium is 95% aqueous ethanol, which is transparent down to about 205 nm. Absolute ethanol cannot be used since it contains traces of benzene, which is 'UV active.' As mentioned above, polar solvents such as ethanol tend to obscure the fine structure of absorption bands. This may be avoided by changing to saturated hydrocarbon solvents, such as cyclohexane, and additionally the range of measurement can be extended down to *ca* 190 nm.

2.5 CHROMOPHORES, AUXOCHROMES AND THE EFFECTS OF CONJUGATION

2.5.1 Chromophores and auxochromes

A group which 'gives rise' to an electronic absorption band is described as a chromophore, but in practice chemists restrict this description to structural units which are unsaturated or aromatic in nature. Atoms or groups bearing non-bonded electrons in orbitals which overlap with the π system of the main chromophore are referred to as auxochromes, since when present they modify the position of λ_{max} of the chromophore through conjugation (see below).

2.5.2 Conjugation

Generally, an extension of the conjugation of a chromophore causes a shift of the position of its absorption maximum to longer wavelengths. Ethene, for example, exhibits $\lambda_{max} = 180$ nm, whereas buta-1,3-diene has $\lambda_{max} = 217$ nm. This bathochromic or red shift comes about because when two isolated chromophores are brought together the orbitals 'mix in' to form a new set of molecular orbitals

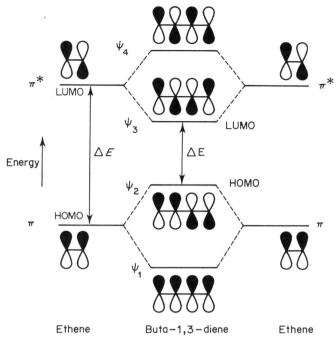

Figure 2.2 Relative energies of ethene and buta-1,3-diene orbitals (schematic).

equal in number to those of the isolated units, but now of different energies.

Thus, for two ethene units linking up to form buta-1, 3-diene the two π and the two π^* (unoccupied) orbitals become four new orbitals ψ_1, ψ_2, ψ_3 and ψ_4 (ψ_3 and ψ_4 being antibonding orbitals). This process is shown schematically in Fig. 2.2, and the important point to note is that the energy gap between the highest occupied orbital (the HOMO) and the lowest unoccupied orbital (the LUMO) in buta-1, 3-diene is smaller than the energy gap between π and π^* in ethene.

It follows, therefore, that if the electronic transition from the HOMO to the LUMO is the dominant feature of the spectrum then it occurs at a longer wavelength (lower energy) in the conjugated system than it does in the unconjugated system.

Similar arguments can be advanced to explain red shifts in the spectra of other molecules in which the chromophore is extended by the attachment of other multiply bonded groups or those bearing non-bonded electron pairs. Another factor which has the same effect is hyperconjugation, although the red shift resulting from the addition of, say, an alkyl group to a double bond is relatively small (see below).

2.6 SELECTED CHROMOPHORIC SYSTEMS

2.6.1 Alkenes and dienes

The absorption maxima of simple alkenes lie at the operational limits of the standard laboratory ultraviolet spectrometer, and only in the case of tetrasubstituted alkenes does λ_{max} approach 200 nm. However, when the double bond forms part of a cyclic system an element of strain may be introduced and in such cases λ_{max} is shifted to longer wavelengths. Compare, for example, the data given below for the two hydrocarbons **2.1** and **2.2**: for the first λ_{max} is close to that of cyclohexene, $\lambda_{max} = 190$ nm ($\varepsilon = 7250$), but for the second not only is the double bond tetrasubstituted, it is also exocyclic to both rings A and C. Together these two effects cause a red shift of 11 nm in the position of λ_{max} relative to that of the parent chromophore.

(2.1)
λ_{max} 193 nm (ε 11 000)

(2.2)
λ_{max} 204 nm (ε 12 000)

Additional conjugated double bonds bring a further shift of $ca + 30$ nm in the position of λ_{max}, but in acyclic systems molecular deformations ensure that beyond six double bonds $\pi-\pi^*$ orbital overlap is progressively less efficient and the increments become smaller. Within cyclic molecules the more rigid geometry

Table 2.1. Fieser–Woodward rules for the prediction of λ_{max} in the spectra of polyenes. Values quoted from Scott (1964) with permission

Moiety	λ_{max} (nm)
Diene chromophore contained in one ring (homoannular)	253
Diene chromophore contained in two rings (heteroannular), or in open chain	214
Increments for:	
alkyl substituent, or ring residue	5
exocyclic double bond*	5
double bond extending the chromophore	30
chlorine, or bromine atom	5
O-alkyl group	6
S-alkyl group	30
O-acyl group	0

*An exocyclic double bond is one that commences at an 'external' apex of a ring system.

of the system 'protects' the chromophore and modifications to it lead to regular shifts in λ_{max}. Indeed, by studying numerous examples taken mainly from the steroid/terpenoid field, Fieser and Woodward (see Fieser and Fieser, 1959) were able to compile a set of rules (see Table 2.1) from which the λ_{max} of cyclic polyenes can be calculated.

Three examples are shown below to illustrate how these calculations can be effected, and it is normally considered that a satisfactory correlation between the observed (obs.) λ_{max} and the calculated (calc.) λ_{max} has been achieved if they differ by not more than 5 nm.

Example 1. Cholesta-3, 5-diene (**2.3**):

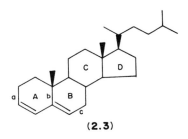

(**2.3**)

In this molecule the diene component is classified as heteroannular because the two double bonds are in different rings, hence λ_{max} of the parent systems is 214 nm. To this must be added 5 nm because one double bond is exocyclic (to ring A), plus three increments of 5 nm since there three ring residues (bonds a, b and c). Hence λ_{max}(calc.) is $214 + 5 + 15 = 234$ nm, which compares favourably with λ_{max}(obs.) = 235 nm.

Example 2. 2-Methyl-4, 4a, 5, 6, 7, 8-hexahydronaphthalene (**2.4**):

(**2.4**)

Here both double bonds are in the same ring, so this is an example of a homoannular diene and λ_{max} of the parent chromophore is 253 nm. One double bond is exocyclic (to ring B), and there are three ring residues (*via* bonds a, c and d). In addition, an alkyl group is joined through bond b to the chromophore. λ_{max}(calc.) $= 253 + 5 + 15 + 5 = 278$ nm; λ_{max}(obs.) = 273 nm.

Example 3. 2-Acetoxy-3, 4, 4a, 6, 7, 8-hexahydro-4a-methylphenanthrene (**2.5**):

(2.5)

In this case there are three possibilities to choose from in selecting the parent diene. By convention the parent is considered to be that which has λ_{max} at the longest wavelength, so here this is the homocyclic diene of ring B. The conjugation of this system is extended by two double bonds (one in ring A and the other in ring C). Three exocyclic double bonds are present (one each to rings A, B and C), there is an acetoxy group (zero increment), and five ring residues are attached through bonds a, b, c, d and e to the overall polyeneic chromophore. λ_{max}(calc.) $= 253 + (2 \times 30) + (3 \times 5) + 0 + (5 \times 5) = 353$ nm; λ_{max}(obs.) $= 355$ nm.

One may easily judge how valuable the technique is, because UV–visible range instrumentation is very much cheaper than that of, say, nuclear magnetic resonance, it is very easy to use and the information obtained provides a reliable method of verifying structural proposals. It is also worth remembering that the measurement of an electronic spectrum requires very small amounts of sample, which may be recovered. The rules do break down, however, when very strained molecules are involved, or when bulky groups about the chromophore prevent good π–π orbital overlap. Diexocyclic dienes such as **2.6** and **2.7**, for example, cannot be analysed by the Fieser–Woodward rules.

(2.6)
λ_{max}(obs.) $= 220$ nm ($\varepsilon = 6\,000$)

(2.7)
λ_{max}(obs.) $= 249$ nm ($\varepsilon = 11\,500$)

2.6.2 Simple carbonyl compounds

Two possibilities for the absorption of ultraviolet radiation occur within the C=O group of aliphatic aldehydes and ketones. Firstly, an electron from the double bond may be promoted into the corresponding π^* antibonding orbital. The related absorption band is observed close to 190 nm ($\varepsilon = 2000$) for solutions in cyclohexane. A second absorption is due to the 'forbidden' n–π^* transition, which occurs at 270–290 nm, and is characterized by its very low intensity ($\varepsilon \leqslant 100$) (see

Section 2.1). The position of this band varies with the solvent, a fact associated with the polar nature of the C=O bond and with the nature of the substituents on the α-carbon atoms. For example, additional alkyl groups at these sites may cause a shift to longer wavelengths since, if they bear C—H bonds, hyperconjugation can occur with the π system of the carbonyl function, thereby extending the overall delocalization of the chromophore and hence the energy of π^*. Unfortunately these effects are not always predictable.

More reliable results are obtained for α-halogenoalkylketones, and in cyclic systems the presence of an axial halogen atom causes a significant red shift. For the corresponding equatorial isomers the shift is much less and can be negative in value (i.e. a blue shift). Thus λ_{max} is often less than that of the parent; compare the data, given below, for the ketones **2.8**, **2.9** and **2.10**. Similar phenomena are observed for other polar substituents, such as hydroxy or acetoxy groups. Thus, within a series of related structures ultraviolet spectroscopy can be used to establish relative configurations.

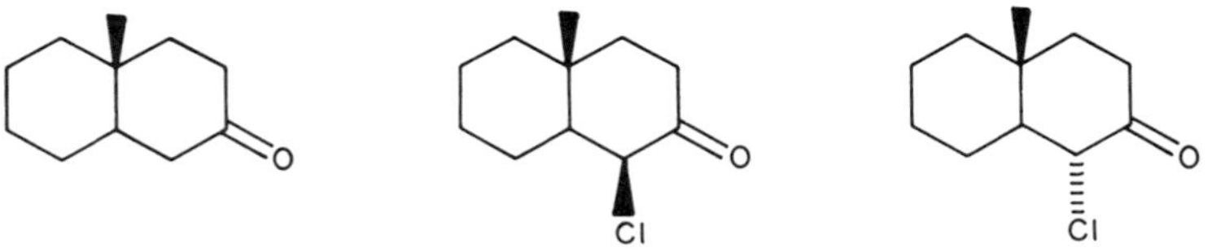

(**2.8**) λ_{max} 287 nm   (**2.9**) λ_{max} 309 nm   (**2.10**) λ_{max} 280 nm

2.6.3 Enones and polyenones

The effect of conjugation in an enone see Fig. 2.3, as for dienes, is to reduce the energy gap between the HOMO and the LUMO of the π system (see page 7). In turn this shifts, to longer wavelengths, the position of the corresponding absorption band, which is commonly referred to as the electron transfer (ET) band to signify its dependence upon the resonance of the π-electrons within the chromophore:

$$C=C-C=O \leftrightarrow {}^+C-C=C-O^-$$

Figure 2.3 Resonance within an enone.

The energies of the orbitals bearing the non-bonded electron pairs of the oxygen atom in an enone are not much changed by conjugation relative to those of the simple carbonyl group but, since the energy of the LUMO is lowered, n–π^* also occurs at longer wavelengths. The intensity of these so-called local excitation

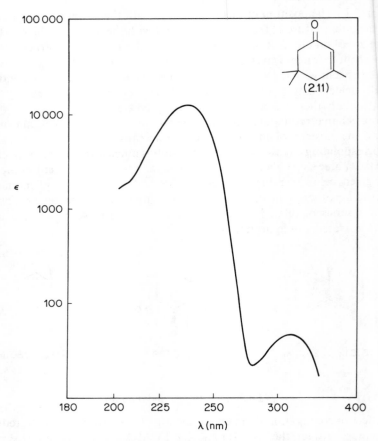

Figure 2.4 Ultraviolet spectrum of 3,5,5-trimethylcyclohexenone recorded on a solution in 95% aqueous ethanol. Data re-plotted, with permission, from *UV Atlas of Organic Compounds*, Butterworths, London, and Verlag Chemie, Verlag Weinheim, 1966–71.

(LE) bands is low since there is poor overlap between the non-bonded-pair orbitals and the LUMO. In the spectrum of 3,5,5-trimethylcyclohexenone (**2.11**) (Fig. 2.4), for example, the LE band occurs at λ_{max} 325 nm ($\varepsilon = 62.5$), whereas the ET band has λ_{max} 234 nm ($\varepsilon = 11\,000$).

Most interest is centred on the positions of absorption of the ET band, and a further set of rules (see Table 2.2) has been drawn up, based on the spectra of many model compounds, to allow the prediction of λ_{max} for enones and polyenones. Once again these rules work well provided that the π systems contributing to the main chromophore maintain good overlap, but in strained or sterically crowded molecules where π–π overlap is impaired the rules break down.

Unlike the spectra of dienes, those of the more polar enones are solvent

ULTRAVIOLET AND VISIBLE SPECTROSCOPY

Table 2.2. Rules for the calculation of λ_{max} of enones and polyenones. Values quoted from Scott (1964) with permission

$$\overset{\delta}{C}=\overset{\gamma}{C}-\overset{\beta}{C}=\overset{\alpha}{C}-C=O$$

Moiety	λ_{max} (nm)
Parent enone when acyclic or in a six-membered ring	215
Parent enone when in a five-membered ring	202
Parent enal (α, β-unsaturated aldehyde)	207
Increments for:	
$\quad$ α-alkyl group, or ring residue	10
$\quad$ β-alkyl group, or ring residue	12
$\quad$ γ- or δ-alkyl group, or ring residue	18
$\quad$ double bond extending the parent chromophore	30
$\quad$ exocyclic double bond	5
$\quad$ homoannular diene	39
$\quad$ α-hydroxy group	35
$\quad$ β-hydroxy group	30
$\quad$ δ-hydroxy group	50
$\quad$ α-methoxy group	35
$\quad$ β-methoxy group	30
$\quad$ γ-methoxy group	17
$\quad$ δ-methoxy group	31
$\quad$ α-chlorine atom	15
$\quad$ β-chlorine atom	12
$\quad$ α-bromine atom	25
$\quad$ β-bromine atom	30
$\quad$ α-, β- or δ-acetoxy group	6
$\quad$ β-amino group (primary, secondary or tertiary)	95

dependent; hence the values in Table 2.2 relate only to measurements taken in 95% ethanol or pure methanol as the solvent. For comparisons with spectra run in other media, correction factors have to be applied (see Table 2.3).

Two examples (**2.12** and **2.13**) serve to illustrate how the rules can be applied to the calculation of λ_{max} of enones and polyenones.

Example 1. Cholest-4, 6-dien-3-one (**2.12**)

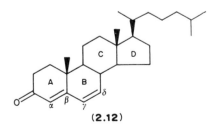

(**2.12**)

Here the enone chromophore in ring A is extended by conjugation with an additional double bond present in ring B (215 + 30 nm), and one double bond is

Table 2.3. Corrections for solvents other than 95% ethanol or methanol

Solvent	Correction (nm)
Chloroform	+ 1
Cyclohexane	+ 11
Diethyl ether	+ 7
Dioxane	+ 5
Hexane	+ 11
Water	− 8

exocyclic to ring B (+ 5 nm). In addition, ring residues are present at the β- and the δ-positions of the dienone (12 + 18 nm), so that in total λ_{max}(calc.) = 280 nm. This compares with λ_{max}(obs.) = 284 nm.

Example 2. Cholest-2, 4-dien-6-one (**2.13**)

(2.13)

The principal difference between this example and the previous one is that here the diene component is located within the same ring (a homoannular system). In addition, one double bond is exocyclic to ring B, and there are ring residues at the α- and δ-positions. The calculation of λ_{max} thus becomes 215 + 30 + 39 + 5 + 10 + 18 = 317 nm, which compares with λ_{max}(obs.) = 314 nm.

In the spectra of complex molecules such as these it is often possible to observe more than one absorption band, but only one is assumed to be the ET band of the polyenone chromophore. The spectrum of the dienone **2.14**, for example, shows two bands, one at λ_{max} 228 nm ($\varepsilon = 11\,600$) and the other at λ_{max} 278 nm ($\varepsilon = 4500$). It is considered that the lower wavelength band is a localized transition associated with the diene component, whereas the longer wavelength absorption reflects an electronic excitation from the fully conjugated chromophoric system. When the absorption maximum of this latter unit is calculated [λ_{max}(calc.) =

(2.14)

281 nm] using the data from Table 2.2 there is a good fit with the experimentally observed figure of 278 nm.

2.6.4 Monosubstituted benzenes

The ultraviolet spectrum of benzene, in hexane as solvent (see Fig. 2.5), exhibits bands at λ_{max} 184 nm ($\varepsilon = 60\,000$), 204 nm ($\varepsilon = 7400$) and 254 nm ($\varepsilon = 204$). These may be considered to be due to the local excitations of the π system, but the low intensity of one of them indicates that it is likely to be a 'forbidden transition' (see Jaffé and Orchin (1962), p. 242). Interestingly, this band shows much fine structure and its position shifts to longer wavelengths as the ring becomes

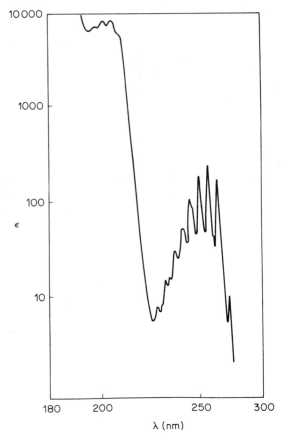

Figure 2.5 Ultraviolet spectrum of benzene recorded as a solution in hexane. Data re-plotted, with permission, from the *UV Atlas of Organic Compounds*, Butterworths, and London, and Verlag Chemie, Weinheim, 1966–71.

more alkylated (compare the data for the three arenes **2.15**, **2.16** and **2.17**).

(**2.15**) λ_{max} 263 nm ($\varepsilon = 300$)   (**2.16**) λ_{max} 266 nm ($\varepsilon = 305$)   (**2.17**) λ_{max} 279 nm ($\varepsilon = 820$)

These shifts are the result of increased conjugation (hyperconjugation), but unfortunately the effects are not predictable and are much influenced by steric factors.

Ring substitution by groups which extend the conjugation of the π system by virtue of extra unsaturation, or by lone pair electron overlap, shifts the position of the high-intensity bands of benzene well into the accessible region of spectral measurement. For example, Table 2.4 lists the principal absorption bands of a number of monosubstituted benzenes, all of which occur in the range 210–300 nm. Naturally, since the origins of these bands are considered to be the result of delocalization of the π-electrons of the ring with those of the substituents, they are designated as ET bands.

It is important to notice the effects of acids and bases on the spectra of arylamines and phenols (see Table 2.4). Protonation of aniline, for example, affords the anilinium cation, in which the lone pair electrons on the nitrogen atom are no longer available to conjugate with the π system of the ring; the spectrum

Table 2.4. Ultraviolet maxima of some monosubstituted benzenes, PhX

X	λ_{max}(ET) (ε)	λ_{max}(LE) (ε)
CN	224 (13 000)	
CHO	242 (14 000), 248 (12 500)	280 (1 400), 289 (1 200)
COMe	243 (12 600)	278 (1 000)
COPh	254 (18 000)	270 (1 700)
COOH	228 (10 000),	279 (500)
COOPh	232 (16 300)	
NO$_2$	269 (7 800)	330 (300)
CH=CH	248 (14 000)	282 (750)
NH$_2$	230 (8 600)	280 (1 430)
NH$_3^+$		254 (160)
NHCOMe	242 (12 000)	
OH	210 (6 200)	270 (1 450)
O$^-$	235 (6 200)	287 (2 600)
OCOMe		258 (290)

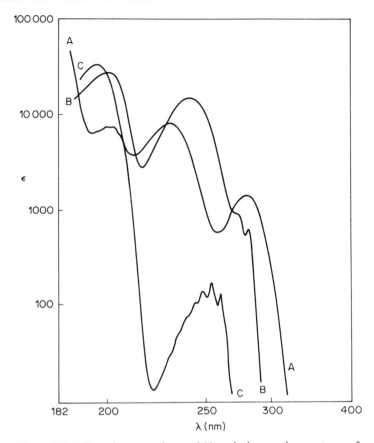

Figure 2.6 Effect of protonation and N-acylation on the spectrum of aniline. A, aniline (hexane); B, acetanilide (hexane); C, anilinium hydrogensulphate (water). Data re-plotted, with permission, from *UV Atlas of Organic Compounds*, Butterworths, London, and Verlag Chemie, Weinheim, 1966–71.

therefore reverts to that of benzene. N-Acylation reduces the delocalization of the nitrogen lone pair electrons with the ring, shifting the ET band to shorter wavelengths (see Fig. 2.6).

Conversely, in phenate salts the release of a full negative charge on the oxygen atom promotes conjugation with the ring and, not only is the ET band of the parent shifted to the red, but also it is usually intensified. This trend is reversed by O-acylation, for now the lone pair electrons on the phenolic oxygen atom are involved in resonance with the carbonyl group and electron transfer into the ring is virtually eliminated (see the last entry in Table 2.4, and Fig. 2.7).

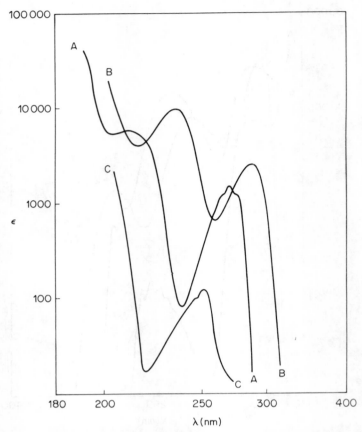

Figure 2.7 Effect of deprotonation and *O*-acylation on the spectrum of phenol. A, Phenol (water); B, sodium phenate (water); C, phenyl acetate (95% ethanol). Data re-plotted, with permission, from *UV Atlas of Organic Compounds*, Butterworths, London, and Verlag Chemie, Weiheim 1966–71.

2.6.5 Disubstituted benzenes

Often the spectra of 1,2- and 1,3-disubstituted benzenes seem to approximate to composites of the spectra of the two appropriate monosubstituted units, and two bands appear at wavelengths which are close to those of the absorption maxima of the two simpler molecules. The same observation also holds for the spectra of some 1,4-disubstituted benzenes, but now only if the two substituents are both electron withdrawing or both electron releasing. When the groups are of different

ULTRAVIOLET AND VISIBLE SPECTROSCOPY

Table 2.5 Absorption spectra of some disubstituted benzenes, XC_6H_4Y

X	Y	Orientation	λ_{max} nm (ε)
OH	OH	1,2-	214 (6 300), 276 (2 300)
OH	OH	1,3-	216 (6 800), 274 (2 000)
OH	OH	1,4-	225 (6 000), 295 (3 100)
OH	NO_2	1,2-	230 (3 900), 278 (6 400), 351 (3 050)
OH	NO_2	1,3-	274 (6 000), 333 (1 950)
OMe	CHO	1,2-	254 (10 000), 322 (4 600)
OMe	CHO	1,3-	252 (8 300), 314 (2 800)
NH_2	NO_2	1,2-	229 (16 000), 275 (5 000), 405 (5 000)
NH_2	NO_2	1,3-	235 (16 000), 373 (1 500)
COOH	COOH	1,2-	230 (9 200), 280 (1 050)
COOH	COOH	1,3-	230 (11 900), 282 (950)

types they act together to extend the conjugation of the π system of the ring and therefore compounds such as 4-nitrophenol exhibit absorption bands at longer wavelengths than either of the two component monosubstituted benzenes. It is likely that in the corresponding *ortho* isomers effective conjugation between the substituents is reduced by steric factors.

Some illustrations which do fit the composite pattern are collected into Table 2.5.

2.6.6 Aromatic carbonyl compounds

Whereas the above observations may owe more to coincidence than to any other factors and have to be treated with some reserve, it is possible to obtain much more reliable information from the spectra of aromatic aldehydes, ketones, acids and esters. As in the aliphatic series, these compounds exhibit weak n–π^* transitions, but now at longer wavelengths of 320–350 nm. More important, they also give rise to an ET band which can be calculated from a set of rules compiled by Scott (see Table 2.6).

A few examples to illustrate the use of the rules are given below.

Example 1. 5-Chloro-2-hydroxybenzaldehyde (**2.18**):

(2.18)

λ_{max} (calc.) = 250 + 0 (m-Cl) + 7 (o-OH) = 257 nm; λ_{max} (obs.) = 255 nm.

Table 2.6. Scott's rules for the calculation of λ_{max}(ET) of aryl carbonyl compounds, ArCOR. Values quoted from Scott (1964) with permission

ArCOR:	
R = H	Parent chromophore = 250 nm
R = alkyl (or ring residue)	Parent chromophore = 246 nm
R = O-alkyl	Parent chromophore = 230 nm
R = OH	Parent chromophore = 230 nm

Substituent	Orientation	Increment (nm)
Alkyl, or ring residue	1,2- or 1,3-	3
Alkyl, or ring residue	1,4-	10
OR (R = H or O-alkyl)	1,2- or 1,3-	7
OR (R = H or O-alkyl)	1,4-	25
O$^-$	1,2-	11
O$^-$	1,3-	20
O$^-$	1,4-	80
Cl	1,2- or 1,3-	0
Cl	1,4-	10
Br	1,2- or 1,3-	2
Br	1,4-	15
NH$_2$	1,2- or 1,3-	13
NH$_2$	1,4-	58
NHCOMe	1,2- or 1,3-	20
NHCOMe	1,4-	45

Example 2. 7-Hydroxyindanone (**2.19**):

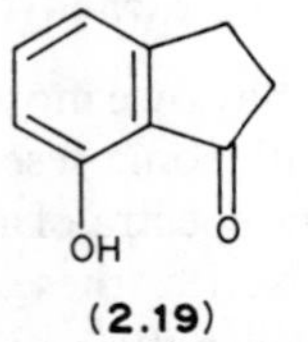

(**2.19**)

λ_{max}(calc.) = 246 + 3 (*o*-ring residue) + 7 (*o*-OH) = 256 nm; λ_{max}(obs.) = 255 nm.

Example 3. 4-Amino-2-methylbenzoic acid (**2.20**):

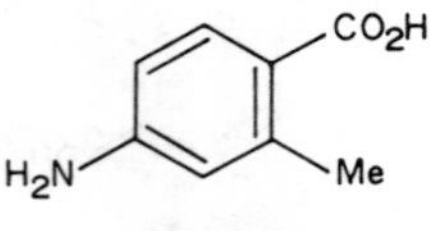

(**2.20**)

λ_{max}(calc.) = 230 + 58 (*p*-NH$_2$) + 3 (*o*-Me) = 291 nm; λ_{max}(obs.) = 288 nm.

2.6.7 Steric effects in aromatic structures

The influence of steric factors on the absorption spectra of aromatic compounds is considerable, and in extreme cases two chromophoric units, formally in conjugation, may only exhibit their own individual absorption maxima. Biphenyl (**2.21**), for example, exhibits a relatively intense ET band which masks the normal LE band of the benzene π system at λ_{max} 254 nm ($\varepsilon = 204$). However, substitution at the *ortho* positions, as in the derivative **2.22**, inhibits conjugation between the two rings and now the spectrum reverts back to that of a simple benzene type.

(**2.21**)
λ_{max} 249 nm ($\varepsilon = 19\,000$)

(**2.22**)
λ_{max} 255 nm ($\varepsilon = 164$)

The usual consequence of less adverse steric hindrance to conjugation is a shift of λ_{max} to longer wavelengths; thus for stilbenes the spectrum of the *cis* isomer **2.23** shows absorption maxima at 224 and 280 nm, whereas that of *trans*-stilbene (**2.24**) has maxima at 228 and 300 nm. In the former compound it is the steric interaction of the *ortho*-hydrogen atoms which impedes the best overlap of the two π clouds.

(**2.23**)
λ_{max} 224 nm ($\varepsilon = 24\,000$), 280 nm ($\varepsilon = 10\,500$)

(**2.24**)
λ_{max} 228 nm ($\varepsilon = 16\,400$), 300 nm ($\varepsilon = 29\,000$)

2.6.8 Polycyclic aromatic and heterocyclic compounds

The shape of the ultraviolet absorption curve for naphthalene (**2.25**) is similar to that of benzene (Fig. 2.5), although the positions of the maxima are shifted to longer wavelengths (see below). The effect of further ring fusions is to cause additional bathochromic shifts, but the extent of these shifts is irregular and depends on the orientation of the aromatic nuclei as this determines the

Table 2.7. Absorption maxima of some polycyclic arenes

Compound	λ_{max} nm(ε)				
Naphthalene	220 (63 000),	275 (56 300),	286 (3980),	312 (2510),	320 (350)
Anthracene	253 (69 000),	323 (6900),	339 (8750),	359 (5910),	375 (5950)
Phenanthrene	250 (84 500),	275 (60 200),	293 (61 300),	330 (610),	346 (700)

resonance energies of the systems; compare the data in Table 2.7 for the polycyclic arenes naphthalene (**2.25**), anthracene (**2.26**) and phenanthrene (**2.27**).

(2.25) (2.26) (2.27)

Groups which extend the delocalization of the π systems of polycyclic arenes cause further bathochromic shifts, but the extent of these shifts vary with the positions of substitution. Thus, for example, the spectra of 1-methylnaphthalene [λ_{max} 223 nm ($\varepsilon = 92900$), 272 nm ($\varepsilon = 8450$), 282 nm ($\varepsilon = 8750$), 293 nm ($\varepsilon = 8120$), 312 nm ($\varepsilon = 790$)] and 2-methylnaphthalene [λ_{max} 224 nm ($\varepsilon = 95400$), 276 nm ($\varepsilon = 8130$), 285 nm ($\varepsilon = 6990$), 305 nm ($\varepsilon = 845$), 320 nm ($\varepsilon = 840$)] show small but obvious differences reflecting changes in the electron distributions in the two molecules.

The spectra of aromatic 6π-electron heterocycles are also influenced by the nature of the resonance within the molecule concerned; thus the unshared pair of electrons on the nitrogen atom of pyridine (**2.28**), for example, lies in an orbital which is orthogonal to the π system of the ring. Its overlap with the π system is minimized and, despite the fact that nitrogen is more electronegative than carbon, the spectrum of pyridine closely resembles that of benzene.

λ_{max} 195 nm ($\varepsilon = 7500$), 251 nm ($\varepsilon = 2000$), 270 nm ($\varepsilon = 450$)

In contrast, the five-membered heterocycle pyrrole (**2.29**) depends on the in-plane conjugation of the electrons on its nitrogen atom with those of the buta-1,3-diene component to make up the sextet of π-electrons needed for aromaticity. The spectrum of pyrrole does not resemble that of benzene.

λ_{max} 210 nm ($\varepsilon = 5100$), 240 nm ($\varepsilon = 3000$) nm

Not surprisingly, the spectra of the bicyclic molecules naphthalene (**2.25**), quinoline (**2.30**) and isoquinoline (**2.31**) are all closely similar, but that of indole (**2.32**) is of a different form.

(**2.30**)   (**2.31**)   (**2.32**)

λ_{max} 226 nm ($\varepsilon = 35\,500$), λ_{max} 218 nm ($\varepsilon = 79\,000$), λ_{max} 220 nm ($\varepsilon = 26\,000$), 270 nm ($\varepsilon = 3\,500$), 300 nm 266 nm ($\varepsilon = 3\,900$), 305 nm 262 nm ($\varepsilon = 6\,310$), 285 nm ($\varepsilon = 2\,000$), 313 nm ($\varepsilon = 2\,500$) ($\varepsilon = 2\,000$), 318 nm ($\varepsilon = 3\,000$) ($\varepsilon = 5\,000$)

BIBLIOGRAPHY

Further information on the origins of electronic spectroscopy can be found in the following books:

Jaffé, J. H. and Orchin, M. (1962), *Theory and Applications of Ultraviolet Spectroscopy*, Wiley, New York.

Mason, S. F. (1963), in 'The electronic absorption spectra of heterocyclic compounds,' Katritzky, A. R. (ed.), *Physical Methods in Heterocyclic Chemistry*, Vol. 2, Academic Press, New York, Ch. 7.

Compilations of spectral data include the following:

Friedel, R. A. and Orchin, M. (1951), *Ultraviolet Spectra of Aromatic Compounds*, Wiley, New York.

Hershenson, H. M. (1956, 1961), *Ultraviolet and Visible Spectra; Index for 1930–1954; Index for 1955–1959*, Academic Press, New York.

Kamlet, M. J. (1960), *Organic Electronic Spectral Data*, Vol. 1 Interscience, New York.

Ungnade, H. E. (1960), *Organic Electronic Spectral Data*, Vol. 2, Interscience, New York.

Wheeler, O. H. and Kaplan, L. (1966), *Organic Electronic Spectral Data*, Vol. 3, Interscience, New York.

Philips, J. P. and Nachod, F. C. (1968), *Organic Electronic Spectral Data*, Vol. 4, Interscience, New York.

Fieser, L. F., and Fieser, M. (1959), *Steroids*, Reinhold, New York.

Scott, A. I. (1964), *Interpretation of the Ultraviolet Spectra of Natural Products*, Pergamon Press, Oxford.

UV Atlas of Organic Compounds, Butterworths, London, and Verlag Chemie, Weinheim, 1966–1971, 5 volumes.

CHAPTER 3

Infrared Spectroscopy

3.1 MEASUREMENT OF SPECTRA

3.1.1 Instrumentation and the presentation of data

Although the infrared spectrum extends from 0.5 to 1000 μm (10^{-6} m), general-purpose infrared spectrometers operate over the range 1–25 μm, which is the region where most organic molecules absorb vibrational energy. The remaining wavelengths can be studied, when necessary, by using more specialized instruments.

In the spectrometer, light from the infrared source is split into a reference beam and an identical beam which passes through the sample. The intensities of the two beams are then compared electronically and the percentage of light transmitted through the sample is recorded against wavelength/wavenumber and plotted directly on the chart paper. The time taken to record the spectrum is thus determined by the rate at which the pen of the chart recorder can scan the working range of wavelengths. This limitation can be overcome by the application of Fourier transform (FT) techniques, which allow high-speed multiple scans of the spectrum to be made electronically. The data obtained from many individual runs are accumulated within a computer and, when required, background noise is subtracted before the spectrum is printed out.

Such FT spectrometers are relatively expensive, but they enable high-definition infrared spectra to be obtained very quickly and require only small amounts of sample; because of these factors the technique is particularly useful in, for example, the analysis of compounds being eluted in rapid succession from chromatographic columns.

The chart paper used with most infrared instruments shows both wavelength and wavenumber, but it is customary to use wavenumber expressed in reciprocal centimetres (cm^{-1}) when discussing infrared data, principally because it is then easier to relate fundamental vibration modes to their overtones and infrared bands to Raman bands.

3.1.2 Sample presentation

It is possible to obtain infrared spectra for vapours, liquids and solids using cells constructed with sodium chloride optics. Such cells are transparent to infrared light over the usual spectral range, but suffer the disadvantage that their surfaces are easily damaged by moisture, either in the sample or in the environment.

Liquids may be examined as films compressed between two sodium chloride plates, and a similar presentation can be used for finely ground solids in the form of a suspension, or mull, in a medium such as liquid paraffin (Nujol) or hexachlorobuta-1,3-diene. Solids may also be studied in solution, often in chloroform or carbon disulphide, but here and in the mulling technique the absorption bands of the solvent or of the medium must be subtracted from the spectra so obtained. A related problem is that certain bands due to the sample may be obscured and a better, although more time-consuming, method is to compress an intimate mixture of the solid and potassium bromide into the form of a thin disc, which is placed directly in the sample beam of the spectrometer. The spectrum so obtained is free from solvent bands, but care has to be taken throughout the operation to exclude moisture.

3.2 ABSORPTION OF INFRARED ENERGY

3.2.1 Modes of vibration

At normal temperatures the bonds of a compound are already vibrating as a result of exchanges with other molecules in the immediate environment. Absorption of energy from an infrared source serves to increase the amplitude of these vibrations, which then rapidly decreases as the excited molecules collide with their less active neighbours. The latter are always in large excess so that the system cannot become saturated with energy. Each vibrational change is quantized, hence the absorption of energy is not continuous but occurs as a series of peaks. The more molecules there are in the light path, the greater is the chance that they may interact with quanta of the appropriate energy; hence low sample concentrations lead to weak spectra.

Two types of vibration are possible within a single bond: it may stretch along its axis, or it may bend (deform). A simple analogy is that of two spheres joined by a weightless spring, and it is easy to see that less energy is required to bend the spring than to stretch it. Not surprisingly, it follows that the bending vibrations of chemical bonds require less energy and occur at lower frequencies than do the corresponding stretching modes.

In polyatomic molecules there are many bonds and numerous vibrational opportunities exist. A structure containing n atoms, where $n > 2$, has in theory $3n - 6$ fundamental vibrational modes; of these, $n - 1$ are stretching oscillations and $2n - 5$ are bending motions.

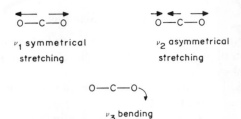

Figure 3.1 Vibrational modes for the linear triatomic molecule carbon dioxide.

Not all molecules, or bonds, absorb infrared energy. The electromagnetic properties of light demand that a particular vibration must produce a fluctuating dipole (and hence a fluctuating electric field). Simple diatomic molecules, such as hydrogen or nitrogen, in the course of vibration do not change their symmetries and hence do not give rise to bands in the infrared spectrum.

Similar effects are noted for the in-plane vibrations of symmetrically substituted multiple bonds. On the other hand, for a polyatomic molecule the lack of

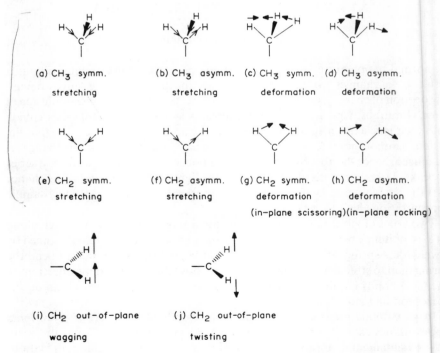

Figure 3.2 Vibrational modes for methyl and methylene groups.

a permanent dipole does not necessarily preclude the absorption of infrared energy. Carbon dioxide, for example, is a symmetrical linear molecule, but should one bond vibrate then an asymmetric diposition of positive and negative centres is achieved. A dipole is momentarily induced and the conditions for the absorption of electromagnetic radiation are fulfilled.

Possible vibrational oscillations for carbon dioxide are illustrated in Fig. 3.1. Of the two stretching modes v_1 and v_2, only the latter is 'infrared active' for the reasons detailed above; the third mode, v_3, is, of course, a bending deformation (there are in fact two degenerate bending modes, the one shown and another identical perturbation which operates at right-angles to it).

More complicated examples are provided by the carbon—hydrogen bonds of alkyl groups. For example, the methyl group has four vibrational modes shown schematically in Fig. 3.2a–d, whereas the methylene group has an additional two oscillations arising from the out-of-plane 'wagging' and 'twisting' deformations i and j.

3.2.2 Vibrational coupling and overtones

When the vibrational energies of two adjacent bonds are closely similar they may interact mechanically and couple together. If this interaction is strong enough, the original fundamental bands are lost and new bands are produced. Thus, whereas an isolated C—H bond has only one stretching frequency, the vibrations of the two bonds of a methylene group couple together to afford the symmetrical and asymmetric stretching combinations discussed above, and illustrated in Fig. 3.2a and b. More complex situations arise when the coupling interactions are weak for then both 'old' and 'new' bands may be observed. Additional bands in the spectrum may be due to overtones, or harmonics ($2 \times v$ or $3 \times v$) of the fundamental band (v), and it is also possible that coupling may occur between the overtones and suitable fundamental vibrations. This last phenomenon is called Fermi resonance after its discoverer Enrico Fermi.

3.2.3 Absorption frequency and intensity

Ideally, the stretching motion of a bond obeys Hooke's law, that is, the degree of stretching is proportional to the restoring force acting on the system. The frequency v is given by the expression

$$v = \frac{1}{2\pi c} \sqrt{\frac{k}{\mu}}$$

where c is the velocity of light, k is the force constant (related to the strength of the bond) μ represents the reduced mass of the bond calculated as $m_1 m_2/(m_1 + m_2)$, m_1 and m_2 being the masses of the two atoms forming the bond. From this expression it follows that, in the absence of other factors, the vibrational frequency of a bond increases as its strength increases and also as the reduced mass diminishes.

The measurement of the absolute intensity of a band does not have the same importance in infrared spectroscopy as it does in ultraviolet–visible spectroscopy (see Section 2.1.2), and normally no attempt is made to measure it. Bands are simply classified in relative terms: strong, medium or weak.

In general, the greater the dipole moment within a bond or group, the more intense is the associated absorption; thus a polar function, such as a carbonyl group, gives rise to a band which dominates the spectrum of the parent compound. A notable exception, however, is the $C\equiv N$ group, which exhibits only a weak absorption band.

3.3 PRACTICAL USES OF INFRARED SPECTROSCOPY

There are two main uses of infrared spectroscopy as an analytical tool: (a) the recognition of particular functional groups within molecules and (b) the identification of compounds by comparison of their spectra with those of authentic samples.

For most compounds, the form of the infrared spectrum is unique, especially in the range $1350-750\,cm^{-1}$, which is sometimes called the 'fingerprint region,' and since most organic laboratories have catalogues of reference spectra, either compiled in-house or purchased commercially, it is a relatively simple matter to identify routine samples.

Similarly, functional groups in isolation, or in combination, give rise to characteristic bands in the spectrum of an unknown compound and as such can be recognized, so it is possible to differentiate between, say, an aldehyde and a ketone or between an amine and an amide by reference to data obtained from model compounds. However, the infrared spectra of all but the simplest molecules are very complex and it is unrealistic to expect to assign more than 10% of the bands present.

The most important regions of the spectrum are usually from 4000 to $2500\,cm^{-1}$ and from 2000 to $1500\,cm^{-1}$. In the former range hydrogen—heteroatom bond stretching absorptions occur, and in the second the vibrations of multiply bonded functions, such as alkene and carbonyl double bonds, are observed. Reference has already beem made to the fingerprint region of the spectrum and it is here that many of the deformation bands occur. Apart from being the 'hall mark' of the the individual compound, some groups exhibit characteristic bands in this region, and from 1000 to $750\,cm^{-1}$, for example, it is possible to use the patterns of the bands to deduce the nature of alkene substitutions or the orientation of substituents on aromatic rings.

A correlation chart for the absorption bands of the commonly encountered functional groups and compounds types is provided in Fig. 3.3, and tables of data

INFRARED SPECTROSCOPY

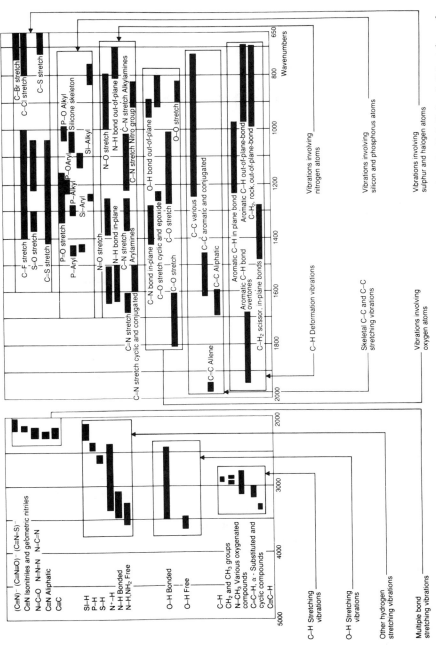

Figure 3.3 Correlation chart for the IR bands of organic molecules. Reprinted with permission from R. C. J. Osland, *Principles and Practices of Infrared Spectroscopy*, Pye Unicam, Cambridge, 1985, p. 17.

and discussions of those factors which affect the appearance or the positions of these bands will be found later in this chapter.

The treatment is not exhaustive and reference is made in the bibliography at the end of the chapter to catalogues of data and more specialized textbooks which will prove helpful when tackling very difficult problems.

3.4 PROBLEM SOLVING

A systematic approach to problem solving is always recommended and adherence to the following points will avoid most pitfalls.

(a) Note the sampling conditions, and make allowance for bands due to solvents, etc.
(b) Start by examining the hydrogen—heteroatom stretching region and cross-check any assignments made by reference to other regions of the spectrum where the associated deformation bands are to be expected.
(c) Study the region 1900–1500 cm^{-1} to ascertain whether any polar multiply bonded groups are present (note that some of these groups also afford diagnostic bands in the deformation regions).
(d) Check for weak bands, such as those due to alkynes and nitriles, or the summation bands of arenes (see Section 3.5). If necessary, re-run the spectrum at higher sample concentrations [see (f) below].
(e) Negative evidence is just as important as positive evidence, but make allowance for the possibility that conjugation between functional groups can lead to a shift in band positions outside of the ranges where the isolated systems absorb.
(f) Finally, if you are running the spectrum yourself, check the calibration of the instrument; often this requires cross-checking against a standard such as polystyrene (refer to the instrument manual).

3.5 THE INFRARED SPECTRA OF ORGANIC COMPOUNDS

3.5.1 Alkanes and alkylated compounds

The C—H stretching bands of saturated hydrocarbons, with the notable exception of cyclopropanes, lie just *below* 3000 cm^{-1}, and this allows a ready differentiation between alkanes on the one hand and unsubstituted alkenes and arenes on the other (the last two exhibit C—H stretching bands *above* 3000 cm^{-1}); compare the spectra of octane (Fig. 3.4) and benzene (Fig. 3.8).

The origins of the fundamental stretching modes of methyl and methylene groups have already been discussed (see Section 3.2.1), and with a high-resolution instrument it is possible to discern both the symmetrical and the asymmetric C—H oscillations of these groups. The methine C—H stretching band is also found below 3000 cm^{-1}, but it tends to be weak.

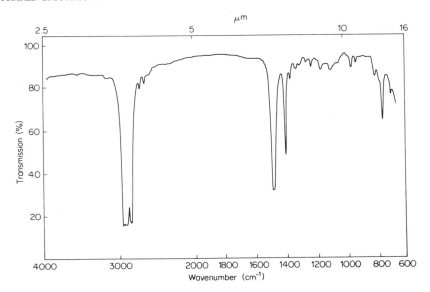

Figure 3.4 The infrared spectrum of octane (liquid phase).

In the case of alkyl halides the influence of the halogen atom is to shift the stretching bands of the adjacent methyl or methylene group above $3000\,\text{cm}^{-1}$; in addition, polar substituents often reduce the intensity of C—H stretching vibrations, but in the deformation region the reverse may be true. For example, in the spectrum of an acetylated compound the symmetric C—H deformation of the methyl group, normally found at $ca\ 1375\,\text{cm}^{-1}$, is intensified and shifted to $1360\,\text{cm}^{-1}$. A similar effect is noted for the symmetric deformation of a methylene group when that group is bound to a polar function. Methyl asymmetric and methylene scissoring oscillations are not normally resolved, but the rocking vibration of the latter group, usually weak, is reinforced as the chain length increases. Thus, when four adjacent methylene groups are present a band of medium intensity is exhibited at $ca\ 750\,\text{cm}^{-1}$ (see Table 3.1).

3.5.2 Alkenes and alkenyl derivatives

Alkenes show bands due to both C—H and C=C stretching modes in their infrared spectra. Monosubstituted alkenes, for example, exhibit three prominent C—H bands: one due to the lone C—H bond and two arising from the symmetrical and asymmetric oscillations of the geminal C—H bonds. These are shown schematically as modes (a) (b) and (c) in Fig. 3.5. The asymmetric band is useful diagnostically as it lies at $ca\ 3100\,\text{cm}^{-1}$, well clear of other aliphatic C—H bands (NB the C—H stretching bands of arenes absorb here), but the

Table 3.1. Infrared bands for alkyl groups

Type	Range (cm^{-1})	Comments
C—H stretch:		
Cyclopropanes	3100–3050, 3035–2995	High wavenumbers make the recognition of cyclopropanes easy
CH$_3$ asymm.	2975–2950	
CH$_2$ asymm.	2940–2915	These bands are only separable at high resolution
CH$_3$ symm.	2885–2860	
CH$_2$ symm.	2870–2845	
CH	2900–2880	Weak
CHO	2900–2800, 2780–2680	Usually a doublet, the lower wavelength band being the weaker
◁, RN◁	3050–3000	
OCH$_3$	2850–2815	
OCH$_2$R	2880–2835	
OCH$_2$O	2880–2835	
OCH(R)O	2820	Weak and variable
NCH$_3$	2820–2760	Amines only; amides do not exhibit this band
CH$_3$Hal.	3058–3005	Iodides at higher and fluorides at lower wavenumbers
CH$_3$CO	3000–2900	
RCH$_2$CO	2925–2850	
C—H deformations:		
CH$_2$ symm. (scissors)	1475–1450	Medium-intensity bands not usually resolved
CH$_3$ asymm.	1470–1430	
CH$_3$ symm.	1395–1365	Often a doublet in compounds bearing geminal methyl groups. C(CH$_3$) units may give rise to triplets, but N(CH$_3$) groups do not absorb in this region
R$_2$CH	~1340	Weak
CH$_3$CO symm.	1385–1360 ⎫	Bands intensified relative to those of simple alkanes
RCH$_2$CO symm.	1435–1405 ⎭	
CH$_3$S	1330–1285	
CH$_3$SO$_2$	1325–1310	
CH$_3$P	1310–1280	
CH$_3$Si	1275–1260	Strong and sharp
Skeletal vibrations:		
RC(CH$_3$)$_3$	1250–1200 ⎫	Broad (unresolved doublet)
R$_2$C(CH$_3$)$_2$	1200–1190 ⎭	Often a doublet
RCH(CH$_3$)$_2$	1175–1140	Doublet
		All the above skeletal bands are subject to variation when polar groups are nearby

Table 3.1. (*Contd.*)

Type	Range (cm^{-1})	Comments
Cycloalkanes	1060–800	Multiple bands, not reliable
Cyclopropanes	1020–1000	
(CH$_3$)$_2$SiR$_2$	855–800	
(CH$_3$)$_3$SiR	840–765	
CH$_3$SiR$_3$	770–760	
CH$_2$ rocking	810–720	Medium intensity, often a doublet. For (CH$_2$)$_n$ where $n \geqslant 4$, $\nu_{max} = 720$ cm^{-1}

symmetrical band occurs at 2975 cm^{-1} and will be masked if alkyl groups are present. The spectra of alkenes may also contain deformation bands such as in-plane and out-of-plane deformation bands analogous to those described for the methylene group (see Fig. 3.2).

Various in-plane absorptions, which occur over the range 1450–1200 cm^{-1}, are weak and of little value, but out-of-plane deformations exhibited at 990–665 cm^{-1} can be used to differentiate between various types of alkenes (see Table 3.2 for details). Opportunities for C—H vibration obviously diminish as the degree of substitution about the double bond increases and symmetrically substituted alkenes are 'inactive' in the infrared region.

Cis-disubstituted alkenes normally show a C=C stretching band at 1660 cm^{-1}, *ca* 15 cm^{-1} lower than the corresponding *trans* isomers, and the latter usually exhibit an out-of-plane deformation band at *ca* 950 cm^{-1} which is absent from the spectra of the *cis* isomers.

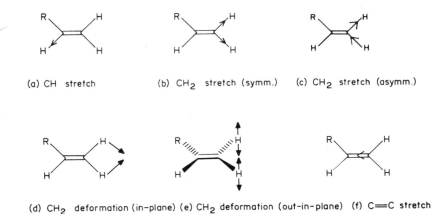

Figure 3.5 Major stretching and deformation modes for monosubstituted alkenes.

Table 3.2. Infrared bands for alkenyl groups

Type	Range (cm^{-1})	Comments
C—H stretch:		
R$_2$C=CH$_2$ asymm.	3095–3075	Multiple bands may occur
R$_2$C=CHR	3045–3010	Differentiation between *cis* and *trans* isomers not possible
C—H deformations		
RCH=CH$_2$	990, 910	
R,RC=CH$_2$	~890	
R,RC=CHR	840–790	
RCH=$\overset{t}{=}$CHR	~950	Whereas the position of the *trans* band can usually be relied upon that of the *cis* cannot
RCH=$\overset{c}{=}$CHR	730–665	
C=C stretch:		
RCH=$\overset{t}{=}$CHR	~1675	Strong–medium band, reliable in acyclic and strain-free systems
RCH=$\overset{c}{=}$CHR	~1660	
RCH=CR$_1$R$_2$	~1670	
R$_2$C=CH$_2$	~1650	
RCH=CH$_2$	~1640	
△ (with exo =CH$_2$)	1780	
△ (with ring C=C)	1640	Monosubstitution at the sp^2 hybridized carbon atoms in these systems by alkyl groups causes shifts in ν_{max} of +10 cm^{-1}. Where disubstitution is possible a further, but smaller, increase in wave number is observed
☐ (with exo =CH$_2$)	1680	
☐ (with ring C=C)	1610	
⬠ (with exo =CH$_2$)	1660	
C=C—C=C	1645–1600	Influenced by geometry of the conjugated system. Intensity usually greater than for a simple alkene
C=C—C=O	1660–1580	
C=C—(C=C)$_n$	1650–1580	Often more than one band, when n is large bands merge into one broad absorption
ArC=C	~1630	Varies with substition pattern of the ring, most influenced by groups at *ortho* and *para* positions

Substitution about a double bond by halogen atoms causes a change in the position of the C=C absorption band relative to that of the hydrocarbon parent; fluorine, for example, induces a positive shift of *ca* 25 cm^{-1}, but chloride, bromine and iodine have the reverse effect. It is noteworthy that *trans*-dihalogenated alkenes, for reasons of symmetry, do not show C=C stretching absorptions. Conjugation with another double bond has the effect of increasing the single-bond character of the original double bond and consequently the position of the C=C stretching band is shifted to lower wavenumbers (see Fig. 3.6).

$$C=C-C=C \leftrightarrow {}^+C-C=C-C^- \text{ etc.}$$

Figure 3.6 Resonance within a diene.

Extra bands may be observed, equal in number to the double bonds in conjugation, but it is usual that these bands have varying intensities and are due to coupled vibrations. In view of this, they cannot be assigned to individual bonds within the system.

3.5.4 Alkynes and alkynyl dervatives (Table 3.3)

Unsubstituted or monosubstituted alkynes give rise to a sharp C—H stretching band at *ca* 3300 cm^{-1}, whereas other absorptions in this region, such as those of alcohols, amines and the overtones of carbonyl groups, are broad. This is a fortunate occurrence, since the C≡C stretching mode is weak. However, care is still necessary since nitriles and isocyanates also absorb in this region. The C—H deformation band of terminal alkynes is also sharp and occurs at 680–610 cm^{-1}, giving rise to an overtone at 1360–1220 cm^{-1}.

3.5.5 Arenes

The vibrational possibilities for benzene are shown in Fig. 3.7 and, as expected, the actual infrared spectrum (Fig. 3.8) is fairly simple because of the symmetry of

Table 3.3. Infrared bands for alkynyl groups

Type	Range (cm^{-1})	Comments
C—H stretch	3340–3250	Usually strong and sharp; but note that N—H and O—H may overlap
C—H deformation	680–610	Associated overtone at 1400–1200 cm^{-1} is of little value
C≡C stretch:		
C≡CR	2260–2190	Weak; *NB* this band is absent in the spectra of symmetrical alkynes
C≡CH	2140–2100	

the molecule. However, this effect is destroyed by monosubstitution and the spectra of such derivatives are more complex. Obviously, similar arguments apply to an understanding of the spectra of polysubstituted arenes; for example symmetrically substituted compounds give rise to spectra containing fewer bands than in those of their unsymmetric analogues.

Usually the C—H stretching oscillations of arenes appear as a series of

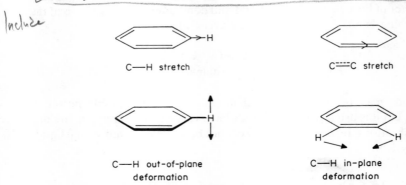

Figure 3.7 Stretching and deformation modes of the benzene ring.

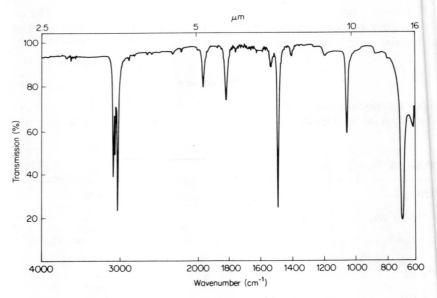

Figure 3.8 Infrared spectrum of benzene (liquid film).

INFRARED SPECTROSCOPY

Table 3.4. Infrared bands for arenes

Type	Range (cm^{-1})	Comments
C—H stretch:		
Ar—H	3100–3010	Multiple bands, medium–weak intensity. Overlap with C—H stretching bands of alkenes
C—H deformations:		
Ar—H in-plane	1250–950	
Ar—H out of plane:		Multiple bands useful for determining the substitution patterns of simple arenes, but unreliable when polar groups are present which extend the conjugation of the molecule. Best used in combination with summation band data (see below)
monosubstituted	~900, 770–730, 720–680	
1,2-disubstituted	780–730	
1,3-	900–860, 865–810,* 810–750, 720–680	
1,4-	830–790	
1,2,3-trisubstituted	800–760, 720–680	
1,2,4-	900–860, 830–790	
1,3,5-	900–840, 850–800, 730–675*	
1,2,3,4-tetrasubstituted	830–790	
Ar—H summation bands	2000–1650	Weak, but observed at high sample concentration. Sometimes used to indicate substitution patterns (see text and Fig. 3.9), but not always reliable
C—C Stretch:		
ArC—C skeletal	1625–1580	Medium–weak, often two bands. Can overlap with C=C stretching band of alkenes, C=N of imines, and C=O bands of enones or vinylogous amides
	1530–1475	Medium, usually two bands which overlap with some alkyl deformation bands

*Not always observed.

medium-intensity bands at 3100–3000 cm^{-1} and normally two or three are exhibited, but a maximum of five is possible.

Up to four skeletal vibrations, due to the ring carbon—carbon bonds, are allowed, and these appear as two pairs of doublets in the ranges 1625–1575 and 1525–1475 cm^{-1}. The first band, often centred at *ca* 1600 cm^{-1} is a useful guide to the recognition of an arene, but it can be overlapped by the absorptions of other groups. Its partner at *ca* 1580 cm^{-1} is normally weak unless ring substituents are present which extend the π system by conjugation. The second pair of doublets are also of differing intensities: that centred at *ca* 1510 cm^{-1} is strongest and this band is often a very obvious feature of the spectra of arenes (see Table 3.4), but that at about 1480 cm^{-1} is weak and frequently overlooked.

Out-of-plane C—H deformations occur below 900 cm^{-1} and the number of adjacent hydrogen atoms present determines the multiplicity and position of these bands. These data can be used to deduce the orientation of substituents around the benzenoid ring, but the presence of halogen atoms and other polar groups renders the analysis unreliable. Moreover, the intensity of some C—H absorptions can be weak resulting in confusion, say, between the identification of monosubstituted and *meta*-disubstituted arenes.

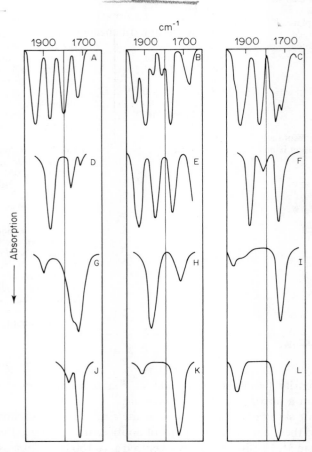

Figure 3.9 Summation bands of arenes. A, Mono-; B, 1,2-di-; C, 1,3-di-; D, 1,4-di-; E, 1,2,3-tri-; F, 1,2,4-tri; G, 1,3,5-tri-; H, 1,2,3,4-tetra-; I, 1,2,3,5-tetra-; J, 1,2,4,5-tetra; K, penta-; L, hexa-. Reprinted with permission from R. C. J. Osland, *Principles and Practices of Infrared Spectroscopy*, Pye Unicam, Cambridge, 1985, p. 13.

Sometimes such problems can be overcome by studying the summation bands (a combination of overtones and coupling interactions) of the C—H deformations. These occur at 2000–1650 cm^{-1} and are best observed at high sample concentrations; here the patterns of the bands, if not their positions, are reasonably independent of the nature of the substituent groups (see Fig. 3.9).

The spectra of polycyclic aromatic hydrocarbons show many similar features to those of benzenes, and in some cases parts of the observed spectrum may resemble a composite of the spectra of the appropriate monocycles. For example, 1,7-disubstituted naphthalenes provide deformation band patterns similar to those of both 1,2,3- and 1,2,5-trisubstituted benzenes.

3.5.6 Aromatic heterocycles

Structurally related aromatic heterocycles normally exhibit similar infrared bands. Thus the spectra of the five-membered heterocycles pyrrole (**3.1**), furan (**3.2**) and thiophene (**3.3**) all show C—H stretching bands in the 3100 cm^{-1} region and multiple skeletal C—C stretching bands at 1610–1515 cm^{-1}. Out-of-plane deformations occur in the range 990–700 cm^{-1} for both furan and thiophene, but at 770–710 cm^{-1} for pyrrole.

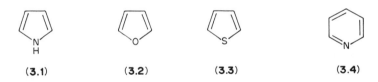

(**3.1**)   (**3.2**)   (**3.3**)   (**3.4**)

The infrared spectrum of pyridine (**3.4**) shows some resemblances to that of benzene, but is more complex since the molecule is unsymmetric. In the free base the C—N stretching band at occurs at 1600 cm^{-1}, but in pyridinium salts the wavenumber shifts to 1620 cm^{-1}.

A major application of infrared spectroscopy in aromatic heterocyclic chemistry is the investigation of tautomeric phenomena. Many amino and hydroxy heteroarenes, for example, can also exist in non-aromatic imino and keto forms, respectively, and in such cases the characteristic bands associated with the different functional groups can be recognized and the ratio of tautomers present assessed. This subject has been well reviewed (Elguero et al., 1976).

3.5.7 Hydroxylated compounds (see Table 3.5)

The infrared spectra of hydroxylated compounds show O—H stretching bands at 3650–3200 cm^{-1} and O—H in-plane and out-of-plane deformations at 1450–1250 and 750–650 cm^{-1}, respectively, plus C—O stretching vibrations at 1210–

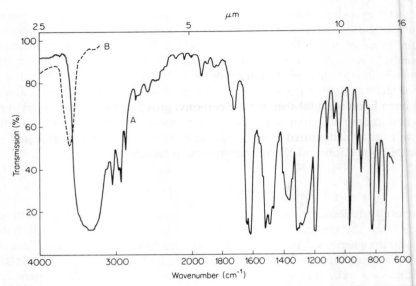

Figure 3.10 Infrared spectrum of *m*-cresol run (A) as a liquid film and (B) as a 5% solution in hexane.

1000 cm^{-1} (in the case of phenols C—O stretching bands are found at $1260-1180 \text{ cm}^{-1}$).

Of these absorptions, that of the O—H stretching mode dominates the spectrum; such a band is often very broad (40–50 cm^{-1}), reflecting the spread of energies resulting from many intermolecular hydrogen bonding opportunities between molecules of the hydroxy compound and between the compound and the solvent. Intermolecular hydrogen bonding is progressively reduced by dilution, and where intermolecular hydrogen bonding is suspected much valuable structural information is gained by re-running the spectrum of a compound at reduced sample concentration (see Fig. 3.10). Similarly, solvent–solute associations are minimized by using solvents of low polarity.

If the steric bulk of substituents adjacent to the hydroxy group prevent intermolecular association, then the O—H stretching band may be sharp. This also applies to the spectra of other molecules in which there is intramolecular hydrogen bonding, for now the energy of the hydrogen bond is more or less defined by the internal structure of the molecule, and a change in the sample concentration has little effect on the appearance of the band. In addition, some dimeric associations are very strong. Aliphatic carboxylic acids, for example, exist almost exclusively in the dimeric form, hence there is little change in the infrared spectra of such compounds even at high dilution. In these acid dimers the

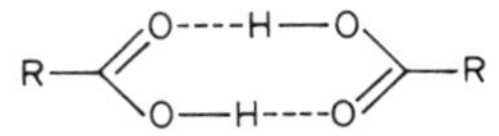

$$\text{R}-\overset{\text{O}----\text{H}-\text{O}}{\underset{\text{O}-\text{H}----\text{O}}{\diamond}}-\text{R}$$

Figure 3.11 Di-molecular association of carboxylic acids.

hydrogen bond is established *via* the carbonyl group as shown in Fig. 3.11 and here, of course, the bond is intermolecular, but in many hydroxy aldehydes, ketones, esters and related compounds the hydroxy and carbonyl groups have suitable orientations to achieve intramolecular hydrogen bonding. Both effects cause the stretching band of the carbonyl group to appear at lower wavenumbers than normal.

The O—H deformations of hydroxylated compounds have little value in structural analyses, but it is possible to distinguish between primary, secondary and tertiary alcohols by the location of their C—O stretching bands, and also to differentiate them from phenols which given rise to similar vibrations, but at higher wavenumbers. A summary of this information is given in Table 3.5.

Table 3.5. Infrared bands of alcohols, phenols and water of crystallization (see also Table 3.1)

Type	Range (cm^{-1})	Comments
O—H stretch:		
ROH unassociated	3650–3580	Sharp band, observed in the spectra of dilute solutions or of vapours. Also for sterically crowded molecules
ROH···HOR dimers	3550–3400	Broad band, shown by some 1,2-disubstituted phenols. Intensity reduced by dilution
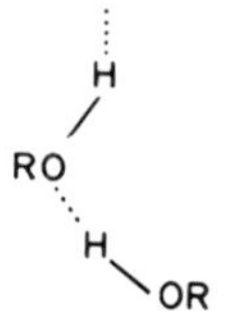 RO⟨H···/H···OR⟩ polymeric	3400–3200	Broad or multiply banded, reverting to unassociated type on dilution
Water of crystallization	3600–3100	This band is associated with a deformation band at 1640–1615 cm^{-1}
O—H deformation:		
ROH in-plane	1450–1250 ⎫	Broad and unreliable
ROH out-of-plane	750–650 ⎭	
C—O stretch:		
ArOH	1260–1180 ⎫	
R$_3$COH	1210–1100 ⎬	Usually strong, but weakened by dilution
R$_2$CHOH	1125–1000	
RCH$_2$OH	1075–1000 ⎭	

3.5.8 Amines

Primary and secondary amines exhibit N—H stretching bands in their infrared spectra. This sets them apart from tertiary amines which, of course, do not possess an NH group. The N—H bond, like the O—H of hydroxy compounds, is susceptible to either intra- or intermolecular hydrogen bonding but, because nitrogen is less electronegative than oxygen, the strengths of the interactions are weaker and the associated bands tend to be less broad.

In dilute solution primary amines given rise to two bands in the 3500–3300 cm^{-1} region resulting from symmetrical and asymmetric stretching modes (see Fig. 3.12, which illustrates the infrared spectrum of aniline).

Secondary amines with only one N—H bond show only one band. Although primary amines exhibit N—H scissoring bands in the 1650–1580 cm^{-1} region, these are weak and, like the N—H deformation and C—N stretching modes, are of little diagnostic value (see Table 3.6).

The salts of amines have different spectra to those of the free bases, because not only are these compounds charged but also, where protonation has occurred, extra N—H bonds are present. As a consequence, extra bands are observed in the N—H stretching range, which is shifted to 3000–2000 cm^{-1}. In addition, these ionic compounds are highly solvated in polar media, causing broadening of the bands. Since amine salts are not readily soluble in non-polar solvents, their infrared spectra are normally recorded on solids (KBr discs).

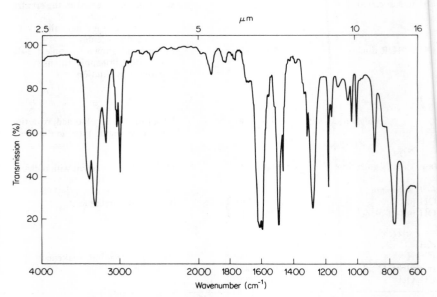

Figure 3.12 Infrared spectrum of aniline (liquid film).

Table 3.6. Infrared bands of amines

Type	Range (cm^{-1})	Comments
N—H stretch:		
RNH_2	3500–3300	Two sharp bands in dilute solution, although in more concentrated media molecular association may be detected by the presence of an additional broader band at *ca* 3200 cm^{-1}. Arylamines also absorb at *ca* 3200 cm^{-1}.
R_2NH	3450–3200	Free amines give rise to one band, but as for RNH_2 association leads to extra bands
$R_2C{=}NH$	3400–3100	One band in dilute solution
$R\overset{+}{N}H_3$	3000–2700	Multiple bands, except at high dilution
$R_2\overset{+}{N}H_2$	3000–2600	Several bands
$R_3\overset{+}{N}H$	2700–1800	Sharp bands
N—H deformations:		
RNH_2	1650–1580 ⎫	Medium–weak intensity bands
R_2NH	1650–1580 ⎭	
Amine salts	1600–1460	Medium–weak band, or bands, RNH_3^+ exhibit symm. bending at *ca* 1300 cm^{-1}
C—N stretch:		
Alkyl—N	1220–1020 ⎫	
Aryl—N	1360–1250 ⎭	Wide range limits value

3.5.9 Aldehydes and ketones

Both aldehydes and ketones give rise to a strong stretching band in the range 1775–1645 cm^{-1}, but only aldehydes exhibit medium–weak absorption at 2830–2690 cm^{-1}, often in the form of a doublet, and due to Fermi resonance (well shown in the spectrum of benzaldehyde in Fig. 3.13).

In common with other carbonyl derivatives, conjugation with double bonds, aryl systems or unsaturated polar groups causes the C=O stretching frequencies of aldehydes and ketones to shift to lower wavenumbers (see Table 3.7). For one double bond this amounts to a decrement of *ca* 30 cm^{-1}. This decrement is increased by further conjugation, but not in a linear fashion as the effect soon tapers off. Hydrogen bonding may also result in a shift to lower wavenumbers (see Section 3.5.7). Conversely, if the carbonyl group is contained within a strained ring system, this leads to a shift to higher wavenumbers; for example, cyclohex-

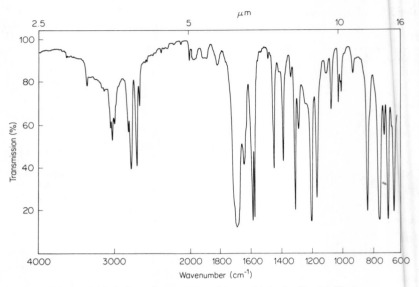

Figure 3.13 Infrared spectrum of benzaldehyde (liquid film).

anone, which is strainless, shows v_{max} at $1715\,\text{cm}^{-1}$, whereas cyclopentanone, which is not, exhibits v_{max} at $1745\,\text{cm}^{-1}$.

α-Halogenation of ketones usually causes a shift to higher wavelengths, but the effect in cyclic systems depends on the dihedral angle between the carbonyl group and the carbon—halogen bond (see p. 11).

3.5.10 Esters, lactones, acyl halides and anhydrides

Carbonyl stretching bands of esters, lactones and acid halides or anhydrides all occur at higher wavenumbers than those of simple aldehydes and ketones. This a consequence of resonance and, particularly, inductive effects which enhance the bond order of the carbonyl group (a detailed discussion of these effects was given by Bellamy, 1975).

In the case of acid fluorides, for example, the carbonyl frequencies lie between $1900-1870\,\text{cm}^{-1}$ and acid anhydrides usually present two bands in the range $1870-1725\,\text{cm}^{-1}$ separated by about $60\,\text{cm}^{-1}$. The positions and intensities of the anhydride bands are altered by the molecular structure: in acyclic anhydrides the higher wavenumber band is the stronger, but the reverse is true for cyclic analogues. The spectra of anhydrides also contain a band at $1050-900\,\text{cm}^{-1}$ resulting from asymmetric stretching of the C—O—C unit.

Ring size has a significant effect on the carbonyl stretching frequencies of lactones (see Table 3.8), and the spectra of both esters and lactones exhibit two

INFRARED SPECTROSCOPY

Table 3.7. Infrared bands of aldehydes and ketones

Type	Range (cm^{-1})	Comments
Aldehydes		
C—H stretch:		
Alkyl—CHO	2830–2810, 2720–2690	Two weak bands
Aryl—CHO	2830–2810, 2750–2720	
C—H deformation:		
RCHO	975–780	Not useful
C=O stretch:		
Alkyl—CHO	1740–1720	Strong, wavenumber reduced by conjugation
Aryl—CHO	1715–1695	Strong, position changed by nature and orientation of ring substituents
Ketones		
C=O stretch:		
(Alkyl)$_2$CO	1725–1705	Strong, altered by conjugation or, in cyclic systems, by ring strain (see below)
Alkaryl—CO	~1690	
(Aryl)$_2$CO	1670–1660	
Conjugated enones	~1675	Possibility of geometric isomers may lead to more than one band
Conjugated dienones	~1665	
cyclobutanone	~1775	
cyclopentanone	~1750	
cyclohexanone	1715	
Large-ring ketones	1715–1705	

strong bands at 1330–1050 cm^{-1} due to the asymmetric and symmetric stretching modes of the C—O bond.

3.5.11 Amides and lactams

The origins of the infrared bands of amides are complex. Thus internal resonance within the amide group imparts much double-bond character to the bond between carbon and nitrogen (see Fig. 3.14). This effect, combined with the fact that the carbon, nitrogen and oxygen atoms have similar masses, brings the

Table 3.8. Infrared bands of esters, lactones, acyl halides, anhydrides and carbamates

Type	Range (cm^{-1})	Comments
C=O stretch:		All bands strong
Esters of aliphatic acids	1750–1735	
Esters of aryl acids	1730–1715	
Alkyl phenolic esters	~1760	Usually strong and reliable
Aryl phenolic esters	~1735	
α, β-Unsaturated acid esters	1730–1710	
α-Keto acid esters	1755–1725	Often only one band
β-Keto acid esters	1750, 1735	Two bands
Lactones (saturated):		
Four-membered	~1840	
Five-membered	~1770	
Six-membered	~1735	
Unsaturated lactones:		
(furanone, C=C away from O)	~1800	
(furanone, C=C adjacent to O)	~1750	Strong and reliable
(pyranone, C=C away from O)	~1760	
(pyranone, C=C adjacent to O)	~1720	
Acyl halides	1815–1785	Range for chlorides and bromides; fluorides 1900–1870 cm^{-1}
Anhydrides:		
If acyclic or in strain-free ring	1840–1800, 1780–1740	Two coupled bands, the first more intense than the second
If in five-membered ring	1870–1830, 1800–1760	Intensity relationship reversed to that of the above
Carbamates	1725–1700	Strong, C—N stretch *ca* 1230 cm^{-1}
C—O stretch:		
Esters and lactones	1330–1050	Strong, two bands
C—O—C stretch:		
Anhydrides:		
Acyclic	~1040	
Cyclic	~920	

Figure 3.14 Resonance within an amide.

stretching frequencies of the carbon—oxygen and carbon—nitrogen bonds within coupling range of one another. New frequencies are created, one of which may couple with the N—H in-the-plane deformation if the amide is either primary or secondary. It is therefore not possible to observe true C=O stretching, N—C—O deformation or N—H deformation bands directly, and instead it is better to refer to the principle absorption bands as amide-I, amide-II, etc.

Despite this, there is a tendency among organic chemists to associate these bands with what they perceive as the major contributor to the coupling process, so that the amide-I band is assumed to be mainly connected with the stretching of the carbonyl group and the amide-II band with the amide N—H in-plane deformation (for other band types, see Table 3.9).

Table 3.9. Infrared bands of amides

Type	Range (cm^{-1})	Comments
N—H stretch:		
CONH$_2$	3500–3100	Two bands in dilute solution; these occur at 3350 and 3180 cm^{-1} in spectra taken at higher concentrations
CONHR	~3450	Transoid
	~3430	Cisoid
C=O, amide-I		
CONH$_2$	1690(free), 1650(assoc.)	
CONHR	1685(free), 1660(assoc.)	Generally strong
CONR$_2$	1650	
Lactams:		
Four-membered	~1750	
Five-membered	~1700	
Larger rings	~1650	
Peptides	1655–1630	
Amide-II, mainly N—H in-plane	1630–1510	Concentration-dependent
Amide-III, C—N stretch, NH deformation	~1400 (primary)	Unreliable
	1290 (secondary)	Higher when associated
Amide-V, N—H out-of-plane	~700	Medium intensity, broad
Amide-IV/VI, NCO deformation	~650	

An increase in double-bond character between carbon and nitrogen leads to the possibility of rotational isomerism, and in many secondary amides and unsymmetrically substituted tertiary amides both E and Z isomers exist as separate compounds at ambient temperature. Naturally, a spectrum of such a mixture shows bands due to both forms.

The presence of N—H bonds in primary and secondary amides ensures a degree of molecular association *via* hydrogen bonding similar to that observed for amines. Therefore, although it is possible to differentiate between these two structural types on the basis of the number of N—H stretching bands, this analysis is only reliable if the spectra are run at high dilution, where intermolecular hydrogen bonding is minimized.

Conjugation of the amide function with unsaturated systems leads to abnormally low wavenumbers for bands in the 'carbonyl region.' Thus the spectra of the vinylous amide rotomers **3.4** and **3.5**, for example, have an amide-I absorption at $1595\,\text{cm}^{-1}$, which is *ca* $40\,\text{cm}^{-1}$ lower than expected for a simple aromatic amide.

(3.4) (3.5)

In the case of lactams the usual influence of ring strain is noted and the 'carbonyl' bands are shifted to higher wavenumber when the amide function is part of a small ring (see Table 3.9).

3.5.12 Carboxylic acids and carboxylate salts

The infrared spectra of carboxylic acids are characterized by a broad O—H stretching absorption, often a series of overlapping bands, from 3300 to $2500\,\text{cm}^{-1}$ (see, for example, the spectrum of heptanoic acid in Fig. 3.15).

In the free state the O—H stretching band is much sharper and occurs at 3650–$3500\,\text{cm}^{-1}$, but carboxylic acids show a strong propensity to form intermolecular associations and in many cases such a band is only observed when the spectra are run at very low sample concentrations (see Section 3.5.7).

The carbonyl stretching band of the carboxylic acid group is strong and very obvious, but its position is also influenced by hydrogen bonding and by conjugation. In addition, two bands resulting from the coupling interactions of C=O and O—H stretching vibrations appear at 1440–1395 and 1320–$1210\,\text{cm}^{-1}$ (see Table 3.10).

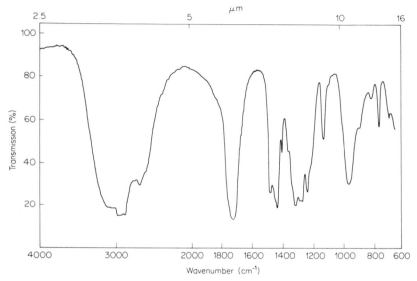

Figure 3.15 Infrared spectrum of heptanoic acid (liquid film).

Table 3.10. Infrared bands of acids

Type	Range (cm^{-1})	Comments
O—H stretch	3550–3500 (free)	Only in dilute solution
	3300–2500 (assoc.)	Very broad band with two maximum at 3000 and 2700 cm^{-1}
O—H deformation	950–900	Broad in the case of dimers
C=O stretch:		
Alkyl—CO$_2$H	1725–1700 ⎫	
α, β-Unsaturated acids	1715–1690 ⎬	These bands are usually broad and *ragged*.
Aryl—CO$_2$H	1700–1680 ⎭	
CO$_2^-$ asymm.	1610–1550	Strong band
CO$_2^-$ symm.	1450–1400	Strong, but easily masked

Carboxylate salts exhibit an asymmetric stretching band at 1610–1550 cm^{-1} and also symmetrical stretching at 1450–1400 cm^{-1}.

3.5.13 Amino acids

Since amino acids normally exist in the zwitterionic form (**3.6**), their infrared spectra display bands due to $^+$NH$_3$ and CO$_2^-$ functions (see Table 3.11), but

Table 3.11. Infrared bands of amino acids

Type	Range (cm^{-1})	Comments
N—H stretch:		
NH_2, NHR	3130–3030, 2800–2000	Broad, almost continuous absorption. Side-band at 2200–2000 cm^{-1}
N—H deformation:		
$\overset{+}{N}H_3$ asymm.	1660–1600	Weak, but sharp
$\overset{+}{N}H_3$ symm.	1550–1480	Medium
NH_2 rocking	1295–1090	
C=O stretch:	1600–1560	In the case of the free acids the C=O stretching band occurs at 1755–1700 cm^{-1}

where the free acid function is present, as in the hydrochloride salts for example, a C=O stretching band is observed at 1755–1700 cm^{-1}.

3.5.14 Azomethines, nitriles, isonitriles and isocyanates

Nitriles, isonitriles and isocyanates all give rise to stretching bands in the range 2275–2115 cm^{-1} (see Table 3.12), but whereas the bands of isonitriles and isocyanates are strong, that of nitriles is weak and can be easily overlooked in some cases.

Table 3.12. Infrared bands of azomethines, nitriles, isonitriles and isocyanates

Type	Range (cm^{-1})	Comments
C=N stretch (azomethines):		
Alkyl—C=N	1690–1640	Medium–weak intensity
Aryl—C=N	1660–1630	
Oximes	1685–1650 (aliphatic)	Oximes show O—H stretching at 3600–3100 cm^{-1}
	1645–1630 (aromatic)	
C≡N stretch (nitriles):		
Alkyl—C≡N	2260–2240	Variable intensity, but weak when conjugated with electron withdrawing groups.
α, β-Unsaturated C≡N	2235–2215	
Aryl—C≡N	2240–2220	
$\overset{+}{N}≡\overset{-}{C}$ stretch (isonitriles)		
Alkyl—$\overset{+}{N}≡\overset{-}{C}$	2145–2135	Variable strong
Aryl—$\overset{+}{N}≡\overset{-}{C}$	2125–2115	
N=C=O (isocyanates)	2275–2240	Intense band due to asymm. stretching. Symm. band at 1350 cm^{-1} is weak

Table 3.13. Infrared bands of sulphur-containing compounds

Type	Range (cm^{-1})	Comments
S—H stretch:		
RSH	2600–2540	Weak
	915–800	
S=O stretch:		
R—SO—R	1060–1020	
RSO—OH	~1090	O—H stretch 2900–2500 cm^{-1} (assoc.)
RSO—OR	~1130	S—O stretch 740–720 and 710–690 cm^{-1}
SO$_2$ stretch:		
R$_2$SO$_2$		
Asymm.	1350–1300	
Symm.	1160–1140	
RSO$_3$H		
Asymm.	1210–1150	
Symm.	1060–1030	
RSO$_3$R		
Asymm.	1420–1330	Strong bands
Symm.	1200–1195	
RSO$_2$NH$_2$		
Asymm.	1370–1330	
Symm.	1180–1160	
RSO$_2$Cl		
Asymm.	1375–1340	
Symm.	1190–1160	
C=S stretch:		
RCSR	1080–1030	Strong, sharp
RCSOR	1210–1080	Strong
RCSNH$_2$	1140–1090	

3.5.15 Sulphur-containing compounds

Thiol groups are much less prone to hydrogen bonding than hydroxy groups and, as a result, S—H stretching bands are less broad than their hydroxy counterparts and in addition they are usually much weaker. Strong absorptions are associated with both sulphoxide (SO) and sulphone (SO$_2$) functions, moreover, the frequencies of these bands are not much changed by their environment, making them very useful structural markers (see Table 3.13).

3.5.16 Ethers

The C—O—C system exhibits a strong asymmetric band at 1290–1030 cm^{-1}, the position of which depends on the nature of the groups flanking the oxygen atom (see Table 3.14). See also Table 3.1 for the C—H stretching frequencies of alkyl ethers.

Table 3.14. Infrared bands of ethers

Type	Range (cm^{-1})	Comments
C—O—C stretch:		
RCH$_2$OCH$_2$R	1150–1085 }	May be two bands
R$_2$CHOCHR$_2$	1170–1115 }	
(epoxide)	1290–1270	cis-Epoxides show an additional band at ca 850 cm^{-1}. For trans isomers the band is at ca 890 cm^{-1}
(oxetane)	~1030, ~980	
(tetrahydrofuran)	~1070, ~915	
(dioxolane)$_n$	950–970	950 cm^{-1} when $n = 1$, 880 cm^{-1} when $n = 2$, 1,3-Dioxanes exhibit a band at ca 800 cm^{-1}
(benzodioxole)	~925	

Table 3.15. Infrared bands of nitro and nitroso compounds

Type	Range (cm^{-1})	Comments
NO$_2$ stretch (nitro):		
Alkyl—NO$_2$		
Asymm.	1565–1545 ⎫	
Symm.	1385–1360 ⎬	
Aryl—NO$_2$		Strong intensity bands
Asymm.	1550–1510 ⎨	
Symm.	1365–1335 ⎭	
NO stretch:		
C-Nitroso	1600–1500 }	
O-Nitroso	1680–1610 }	Strong intensity bands
N-Nitroso	~1450	
C—N stretch:		
Alkyl—NO$_2$	920–830	
Aryl—NO$_2$	860–840	
C—NO	850 (aliphatic), 1100 (aromatic)	

3.5.17 Nitro and nitroso groups

The nitro group shows strong asymmetric and symmetrical bands at 1565–1510 and 1385–1335 cm^{-1}, respectively. Aromatic nitro compounds absorb at lower wavenumbers than their aliphatic analogues, but as a cross-check the former also show a strong stretching band in the 860–840 cm^{-1} range. Aliphatic nitro compounds absorb only weakly in this part of the spectrum (see Table 3.15).

C-Nitroso groups give rise to only one N═O stretching band at 1600–1500 cm^{-1} and a broad C—N stretching absorption at ca 1100 cm^{-1} in the case of aromatic compounds.

BIBLIOGRAPHY

For a comprehensive analysis of the origins and applications of infrared spectroscopy see:

Bellamy, L. J. (1975), The Infrared Spectra of Complex Molecules, 3rd ed., Chapman and Hall, London.

Bellamy, L. J. (1975), *Advances in Infrared Group Frequencies*, Chapman and Hall, London.

There are several excellent collections of infrared spectra, which include the following:

Documentation of Molecular Spectroscopy, Butterworths, London, and Verlag Chemie, Weinheim.

Sadtler Standard Spectra, Heyden, London, 1970.

Pretsch, E., Seibel, J., Simon, W., and Clerc, T. (1983), *Tables of Spectral Data for Structure Determination of Organic Compounds*, Springer-Verlag, Berlin.

Pouchert, C. J. (1986), *The Aldrich Library of FT-Infrared Spectra*, 1st ed., Aldrich Chemical, Milwaukee.

Keller, R. J. (1986). *The Sigma Library of FT-Infrared Spectra*, 1st ed., by Nicolet Instruments, St. Louis.

A detailed appraisal of the use of spectroscopic techniques (including infrared methods) for the investigation of tautomeric phenomena can be found in:

Elguero, J., Marzin, C., Katritzky, A. R., and Linda, P. (1976), *The Tautomerism of Heterocycles, Supplement 1, Advances in Heterocyclic Chemistry*, Academic Press, New York.

A concise introduction to the technique of infrared spectroscopy can be found in:

Osland, R. C. J., *Principles and Practices of Infrared Spectroscopy*, 2nd ed., Pye Unicam, Cambridge.

CHAPTER 4

Nuclear Magnetic Resonance (NMR) Spectroscopy

4.1 INTRODUCTION

NMR spectroscopy is the most important spectroscopic method for the determination of molecular structure and stereochemistry. It is applied across the areas of organic, inorganic, organometallic, biological and medicinal chemistry providing detailed information not only on discrete, low-molecular-weight compounds but also on synthetic and natural polymers and macromolecules. In addition, the technique has many routinely employed applications including those to biosynthetic studies, the direct observation of an increasing number of cellular processes, chemical dynamics and in live organ and whole body imaging. The contents of this chapter, however, will be confined mainly to the determination of structures of organic molecules using proton and carbon-13 NMR spectroscopy.

A great deal of success can be achieved in the interpretation of these spectra without a knowledge of their underlying principles. A 'black-box' approach will therefore initially be adopted, with consideration of the theoretical basis of the subject being made later in the chapter after the reader's interest has, it is hoped, been stimulated by the pleasure gained in the surprising ease with which compounds reveal their identity from 'a few lines on a page.' In order to do this it is now necessary, in discussing the four important parameters from which information is extracted, to use a number of terms which will be defined and explained more fully later.

4.2 THE NMR PARAMETERS

The NMR spectrum is a plot of absorption of radio-frequency radiation against chemical shift (δ) and is often determined on an approximately 10% solution of a test compound in $CDCl_3$. In the spectrum reproduced in Fig. 4.1 an absorption peak appears at 0.0 ppm. This is due to the protons of tetramethylsilane (TMS), a widely used internal reference compound. TMS is convenient since it is chemically inert, produces a sharp singlet absorption (but see Appendix 4.16.4)

NUCLEAR MAGNETIC RESONANCE SPECTROSCOPY

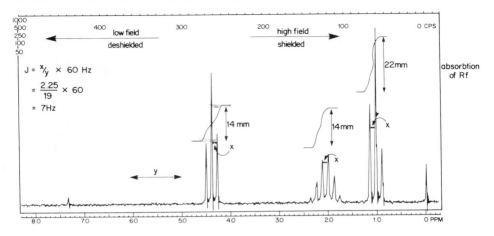

Figure 4.1 Reproduced from *NMR Spectra Catalog* 1962, by permission of Varian Associates. © Varian Associates, 1962.

and resonates at higher field (more negative value) than the protons of most compounds.

1. In Fig. 4.1 three groups of absorptions corresponding to three different sets of protons are present at different *chemical shift* (δ) values (centred at approximately 1.0, 2.0 and 4.4 ppm). The scale represents the shift of the resonance positions downfield from TMS expressed in ppm; δ is a dimensionless constant for a given chemical environment. The assignment of protons to different chemical environments given by reference to their δ values is the first useful parameter of NMR spectroscopy. Three further useful parameters follow.
2. An *integral trace* is run over the spectrum (differential trace), providing a measure of the area under each group of signals. The area under the trace is directly proportional to the number of protons giving rise to that signal. In this case the integral ratio found, on going from high field to low field, is 3:2:2.
3. The high-field and low-field groups have a *multiplicity* of three and the central group of six, i.e. they appear as two triplets and one sextet. The number of nuclei *adjacent* to a given nucleus or group of equivalent nuclei can be estimated from the multiplicity of that signal arising from that nucleus or given set of nuclei.
4. The separations between the individual lines of each group are identical and are identical with the separations in the other groups of protons. This separation is called the *coupling constant* (J) and is measured in hertz (cycles per second). This J value can be used mainly to draw conclusions about

hybridization of the carbons carrying the protons in question and the stereochemical relationships between the coupled protons.

As we interpret each spectrum then we shall draw information from these four parameters, which are now discussed further.

4.3 CHEMICAL SHIFT (δ)

The chemical shift at which a given proton will appear can be predicted using simple empirical equations based on a large body of collected data. Three such equations, and their associated tables of increments, relating to the positions of aliphatic, aromatic and olefinic protons, appear in Appendix 4.16.1. However, in practice there is often no need to use these tables since different proton types appear in well defined regions of the spectrum (see Fig. 4.43). In general, their position is dependent on the electron density about their environment, which in turn relates most importantly to inductive and resonance effects, transmitted through bonds, and anisotropic effects through space.

4.3.1 Inductive effect

The clear relationship existing between the size of chemical shift and the electronegativities (ENs) of adjacent atoms and groups can be seen from Table 4.1.

For multiple substitution of a given carbon atom, the additivity effect is often, but not always, approximately linear. This is illustrated in Table 4.2.

As we infer from acidity measurements of substituted carboxylic acids, the inductive effect decreases rapidly through a singly bonded carbon chain, and this is borne out by the data in Table 4.3.

Table 4.1. Variation of $\delta(CH_3X)$ with the electronegativity of X

X	F	OH	NH_2	Br	I	SH	H
EN	4.0	3.5	3.0	2.8	2.5	2.5	2.1
δ	4.26	3.4	2.5	2.7	2.2	2.1	0.23

Table 4.2. Additivity effect of substituents on δ values

X	Me	Cl	Ph	OMe
CH_3X	0.9	3.05	2.35	3.25
CH_2X_2	—	5.3	4.0	4.5
CHX_3	—	7.25	5.55	5.0

Table 4.3. Variation of δ with distance from OH group in alkanols

$CH_3CH_2CH_2CH_3$	$HOCH_2CH_2CH_3$	$HOCH_2CH_3$	$HOCH_3$
1.3 0.9	1.5 0.9	1.2	3.4
	$HOCH_2CH_2CH_3$	$HOCH(CH_3)_2$	$HOCH_2-$
	1.3 0.9	1.15	3.5
		$HOC(CH_3)_3$	$HOCH-$
		1.2	3.9

A trend similar to but less marked than that observed in Table 4.2 can be detected, for less obvious reasons, with increasing alkyl substitution. This is perhaps explained by small changes in hybridization of the orbitals around the substituted carbon due to increasing crowding, giving RCH_3 (0.9 ppm), R_2CH_2 (1.25 ppm) and R_2CH (1.5 ppm).

4.3.2 Hybridization effect

This may be regarded as another form of inductive effect where the electronegativity of carbon increases with increasing s-character, explaining in part why alkenic hydrogens have a range of about 4.5–7.5 ppm, benzene protons resonate at 7.25 ppm and aldehydic protons at about 9.8 ppm. It is only in part because functional groups containing π systems give rise to both inductive and anisotropic effects.

4.3.3 Anisotropic effects

The π systems of electrons such as those above also give rise to anisotropic effects which can cause both upfield (shielding) and downfield (deshielding) shifts. In general, sp^2 systems, much the most commonly encountered, give rise to shielding for protons lying above or below the plane of the systems and deshielding for those lying in the plane. These effects are illustrated below.

The shielding ($+$) and deshielding ($-$) regions associated with the anisotropy of the various π systems are summarized in Fig. 4.2, where (a) represents sp^2 systems including alkene, carbonyl, imine, aromatic and nitroso and (b) represents the sp systems including alkyne and nitrile.

Examples of this are **4.1** and **4.2**, where the *ortho* protons, falling in the deshielding regions of both the aromatic nucleus and its substituents, appear at 0.5–0.9 ppm to lower field than in benzene; **4.3**, where one proton falls in the deshielding region of the amide group; **4.4**, where the C-4 and C-5 methylene groups are held in the shielding region of the aromatic π-system and have a negative δ value (-1.0 ppm), whereas the C-1 type methylenes appear below 2.0 ppm; and **4.5**, where an example of shielding by the alkene bond is illustrated.

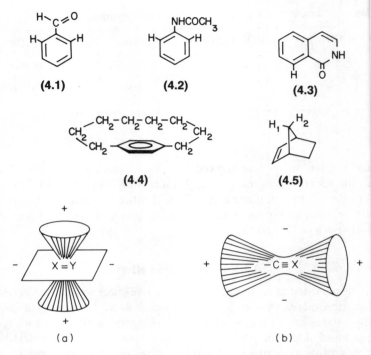

Figure 4.2 Shielding (+) and deshielding (−) regions associated with the anisotropy of the various π systems.

4.3.4 Resonance effects

Although they cannot be readily separated from inductive and hybridization effects, these also influence chemical shift values. Hence the relative shifts of the protons shown in **4.6** and **4.7** can be explained by reference to their charge-separated canonical forms. The operation of resonance effects can be seen in the data in Appendix 4.16.1, especially in the tables concerning the calculation of proton shifts for alkenes and aromatics substituted by powerful −M or +M groups.

4.3.5 Hydrogen bonding effects

The stronger the hydrogen bond in which it is engaged, the more a proton will be deshielded. Carboxylic acids (10–12 δ) provide an example where resonance, inductive and hydrogen bonding effects combine to produce a large downfield shift from TMS. Since the degree of hydrogen bonding is solvent and concentration dependent, the chemical shifts of acidic protons in general are accordingly variable (see Appendix 4.16.2). Such protons can be identified by shaking the sample solution with D_2O and then re-running the spectrum. The acidic protons will exchange with deuterium atoms, which incidentally are NMR active but absorb at a different frequency, and thus disappear from the spectrum, to be replaced by a singlet due to HOD (4.8 δ).

4.3.6

A small set of shift positions which can be easily memorized can be found in Appendix 4.16.1, Fig. 4.43.

4.4 MULTIPLICITY; SPIN–SPIN SPLITTING; $n+1$ Rule

Returning to Fig. 4.1, the spectrum shown may be described by the following, self-evident, shorthand notation. ^{1}H NMR, δ(CDCl$_3$); 1.0 (3H, t, $J = 7$ Hz), 2.0 (2H, sex, $J = 7$ Hz) and 4.4 (2H, t, $J = 7$ Hz). For a given set of protons, their multiplicity will be one more than the number of equivalent protons adjacent to (or splitting) that set. This is a statement of the $n+1$ rule. Where this rule is observed, first-order coupling is said to apply. Thus the sextet of peaks centred at 2.0 ppm has $6 - 1 = 5$ equivalent, adjacent protons. The most obvious part structure which has this characteristic is $^3CH_3\ ^2CH_2\ ^1CH_2-$. If this is correct, then the signal centred at 2.0 ppm must be due to the C-2 protons. The C-1 methylene and the methyl groups are each adjacent to the C-2 methylene and so each appears as a triplet $(2 + 1 = 3)$. The methyl group (integrating to 3H) must then be at 1.0 ppm and the C-3 methylene (integrating to 2H) at 4.4 ppm. The low-field position of the C-3 methylene suggests the structure $CH_3CH_2CH_2X$, where X is an electron-withdrawing group or atom such as NO_2, OR, F or Cl. Notice that X cannot contain a different type of hydrogen since no other signal appears in the spectrum, so that group R could only be propyl. Chemical shift data (see Appendix 4.16.1) point to $X = NO_2$, as also would elemental, mass spectral and infrared analyses. The coupling constant of 7 Hz, the calculation of which is shown in Fig. 4.1, is also a normal value (see Section 4.6 and Appendix 4.16.2) for coupling between vicinal hydrogens in an alkyl chain.

4.5 PASCAL'S TRIANGLE

The relative intensities of the constituent lines of each multiplet are given by the coefficients of the binomial expansion of $(1 + x)^n$ or by Pascal's triangle (Fig. 4.3).

Number of adjacent nuclei	Intensity of lines observed	Abbreviation used
0	1	Singlet s
1	1 1	Doublet d
2	1 2 1	Triplet t
3	1 3 3 1	Quartet q
4	1 4 6 4 1	Pentet p
5	1 5 10 10 5 1	Sextet sex
6	1 6 15 20 15 6 1	Heptet h

Figure 4.3 Pascal's triangle.

Sometimes the low-intensity outside lines of a multiplet, e.g. a heptet, can be 'lost' in the background 'noise' of a spectrum, giving the appearance, in this case, of a pentet. However, inspection of the intensity ratios for the example chosen, i.e. 6:15:20:15:6 as opposed to 1:4:6:4:1, usually resolves the problem.

4.6 COUPLING CONSTANTS (J)

The distance between the lines of a multiplet measured in Hz (cycles per second) is termed the coupling constant, J. As implied, this value is independent of the applied magnetic field and can be read off the ppm scale, which is also calibrated in hertz. In a spectrum run at 60 MHz (60×10^6 Hz), 1 ppm = 60 Hz, at 100 MHz 1 ppm = 100 Hz, and so on. In Fig. 4.1, therefore, one division equals 6 Hz.

4.7 COMMONLY ENCOUNTERED COUPLED SYSTEMS

A number of splitting patterns corresponding to particular part structures become familiar after only a brief acquaintance with proton spectroscopy. Most of the frequently encountered patterns are now listed. A common feature of these is the large $\Delta v/J$ ratio which is necessary if first-order coupling (i.e. adherence to the $n + 1$ rule) is to be observed. Letters well separated from each other in the alphabet are used to describe coupled nuclei well separated in chemical shift; lower field protons are allocated early letters and high-field protons later letters. As the $\Delta v/J$ ratio becomes smaller, so the departure from the intensity ratios indicated by Pascal's triangle becomes greater, the outside lines decreasing and the inside lines increasing in intensity. This phenomenon gives rise to the often seen 'pointing together' of coupled systems as shown by arrows in Fig. 4.4b. As $\Delta v/J$ falls to ca 4 and lower, spectra can sometimes still be analysed as if first-order splitting applies, although extra lines begin to appear and J and δ values cannot be extracted directly from the spectrum.

4.7.1 The AX system

The AX spectrum consists of a pair of doublets ideally as in Fig. 4.4a, but usually seen as in Fig. 4.4b. As the $\Delta v/J$ ratio decreases the system becomes AB as in Fig. 4.4c and ultimately when $\Delta v/J = 0$ the A_2 in Fig. 4.4d results.

NUCLEAR MAGNETIC RESONANCE SPECTROSCOPY

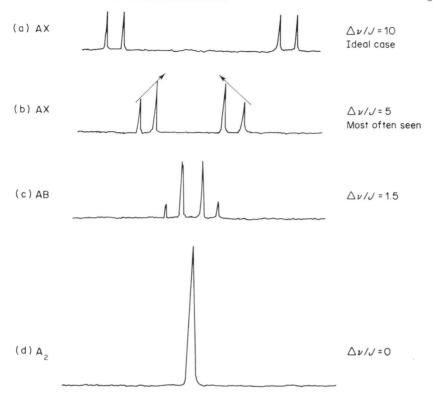

Figure 4.4 Change in the AX spectrum with decrease in the $\Delta v/J$ ratio.

Some examples of the variety of molecular fragments which can give rise to the AX system are given in Table 4.4.

Table 4.4. Molecular fragments that give rise to the AX system

Molecule	δ(ppm) [J(Hz)]	System
(a) (E)-$C_6H_5CH\!=\!CHCOOH$	7.8(1H, d, 16 Hz); 6.5(1H, d, 16 Hz)	AX
(b) p-MeO$C_6H_4CHClCHClCOOMe$ (erythro)	5.1(1H, d, 10.5 Hz); 4.5(1H, d, 10.5 Hz)	AX
(c) 2,4-Dichloro-3-methylacetanilide	8.10(1H, d, 8.5 Hz); 7.35(1H, d, 8.5 Hz)	AX
(d) $Cl_2CHCH(OEt)_2$	5.55(1H, d, 5.5 Hz); 4.6(1H, d, 5.5 Hz)	AX
(e) 5-Methylfurfural	7.3(1H, d, 5.5 Hz); 6.6(1H, d, 5.5 Hz)	AX
(f) Cl(CN)C$=$CH$_2$	6.2(1H, d, 3 Hz); 6.1(1H, d, 3 Hz)	AB
(g) $CH_3CH(OCH_AH_BC_6H_5)_2$	5.2(2H, d, 12 Hz); 5.1(2H, d, 12 Hz)	AB
(h) ArCH$_2$N\...NCH$_A$H$_x$Ar	5.55(1H, d, 15 Hz); 4.05(1H, d, 15 Hz)	AX

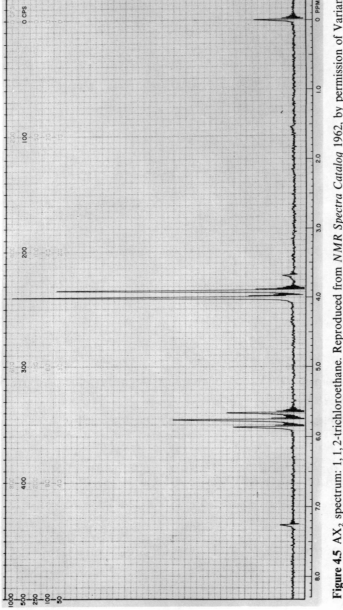

Figure 4.5 AX₂ spectrum: 1,1,2-trichloroethane. Reproduced from *NMR Spectra Catalog* 1962, by permission of Varian Associates. © Varian Associates, 1962.

NUCLEAR MAGNETIC RESONANCE SPECTROSCOPY

Table 4.5. Molecules that give rise to the AX_2 system

Molecule	δ(ppm) [J (Hz)]	System
(a) $CHCl_2CH_2Cl$	5.8 (1H, t, 6.0 Hz); 4.0 (2H, d, 6.0 Hz)	AX_2
(b) $C_6H_5CH_2CHO$	9.7 (1H, t, 3.0 Hz); 3.7 (2H, d, 3.0 Hz)	AX_2
(c) $CH_2{=}C{=}CH(OEt)$	5.7 (1H, t, 6.5 Hz); 4.9 (2H, d, 6.5 Hz)	AX_2
(d) $HC{\equiv}CCH_2OH$	4.3 (2H, d, 2.0 Hz); 2.5 (1H, t, 2.0 Hz)	A_2X
(e) $BrCH_2CH(OEt)$	4.7 (1H, t, 6.0 Hz); 3.4 (2H, d, 6.0 Hz)	AX_2
(f) 2,6-Dinitroaniline	8.6 (2H, d, 8.0 Hz); 6.8 (1H, t, 8.0 Hz)	A_2X

4.7.2 The AX_2 and related systems

4.7.2.1 The AX_2 system

The AX_2 system appears as a doublet and a triplet, as shown in Fig. 4.5.

Examples of the variety of organic molecules from which this pattern arises are shown in Table 4.5.

4.7.2.2 The A_2B system

When small $\Delta v/J$ ratios occur, the spectrum becomes A_2B, where up to nine lines are observed, one or two of which are often very weak or absent as in the spectrum shown in Fig. 4.6.

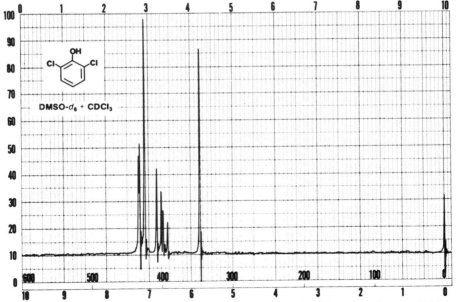

Figure 4.6 A_2B spectrum: 2,6-dichlorophenol. Reprinted with permission of Aldrich Chemical Co., Inc.

4.7.2.3 The AMX system

Both the AMX and the ABX spin systems are frequently encountered, but the ABX system (see Section 4.12.1) usually becomes AMX when high-field (270 or 400 MHz) spectra are run. The AMX pattern is easily recognized by its three groups of four lines. The spectrum of the vinylic protons of styrene, an example of such, can be explained as follows using the 'family-tree analysis' in Fig. 4.7.

(1) Represents the chemical shift positions of the protons without coupling and in (2), (3) and (4) the effect on the spectrum of introducing successively the three couplings is depicted until, in sum, (5) illustrates the full twelve-line spectrum. Proton H_M is assigned the middle δ value as, for maximum overlap of the two π systems, the molecule tends to planarity, thus placing H_M more in the deshielding region of the benzene nucleus than H_X. A second AMX spectrum, due to the alkyne HC≡CCH=CHOMe, is shown in Fig. 4.8.

If any two of the three J values are equal, then the spectrum simplifies to eleven lines. Thus, in the high-field (400 MHz) spectrum of $CH_AH_MBrCH_XBrCOOH$, J_{AM} may be taken as equal to J_{AX} and so H_M appears as a triplet in Fig. 4.9. Where

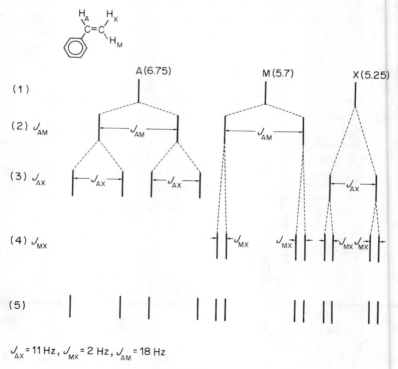

$J_{AX} = 11$ Hz, $J_{MX} = 2$ Hz, $J_{AM} = 18$ Hz

Figure 4.7 The AMX 'family tree.'

NUCLEAR MAGNETIC RESONANCE SPECTROSCOPY

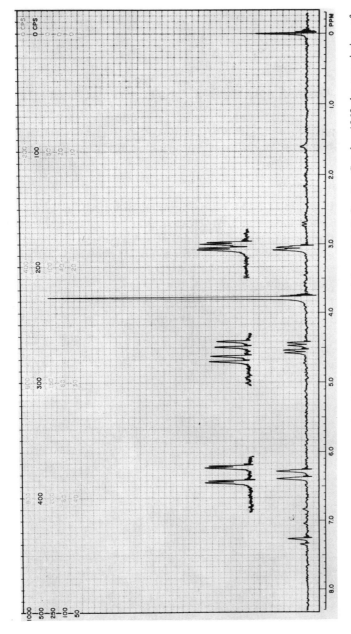

Figure 4.8 AMX spectrum: 1-methoxybut-1-en-3-yne. Reproduced from *NMR Spectra Catalog* 1962, by permission of Varian Associates. © Varian Associates, 1962.

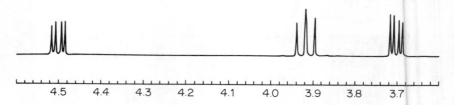

Figure 4.9 AMX spectrum: 2,3-dibromopropionic acid.

one of the coupling constants is zero or close to zero, frequently the case in 1, 2, 4-trisubstituted benzenes, an eight-line spectrum as in Fig. 4.21 is seen.

4.7.2.4 The ABX system

At low field strengths (60–100 MHz), the germinal protons of the CH_AH_B—CH_X part structure often have small $\Delta v/J$ ratios. Here the system, irrespective of whether the X portion is to high or low field of the AB portion, is termed ABX and not AWX as consistency would require for the latter possibility. An example of the ABX pattern is shown in Fig. 4.10, which (see Fig. 4.9) becomes AMX at 400 MHz and further examples are listed in Table 4.6.

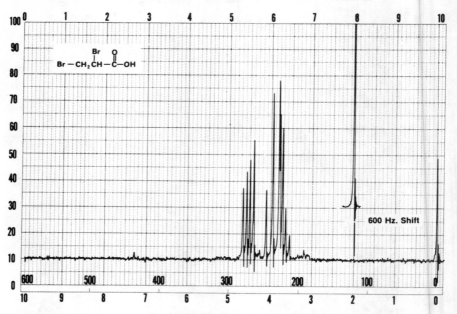

Figure 4.10 ABX spectrum: 2,3-dibromopropionic acid. Reprinted with permission of Aldrich Chemical Co., Inc.

NUCLEAR MAGNETIC RESONANCE SPECTROSCOPY

Table 4.6. Molecules that give rise to the ABX system

Molecule	δ(ppm)[J(Hz)]	System
(a) 2-Nitrocinnamaldehyde	9.8(1H, d, 8.0 Hz); 8.2(1H, d, 16.5 Hz); 6.7(1H, d, 8.0 and 16.5 Hz)	AMX ($J_{AM} = 0$)
(b) Furan-2-carboxaldehyde	7.5(1H, d, 2.5 Hz); 7.30(1H, d, 6.0 Hz); 6.65(1H, d, 2.5 and 6.0 Hz)	AMX ($J_{AM} = 0$)
(c) 4-Chlorostyrene	6.75(1H, dd, 11.5 and 18.0 Hz); 5.65(1H, dd, 2.0 and 18.0 Hz); 5.2(1H, dd, 2.0 and 11.5 Hz)	AMX
(d) Styrene oxide	3.8(1H, dd, 3.0 and 4.0 Hz); 2.90(1H, dd, 4.0 and 5.0 Hz); 2.75(1H, dd, 3.0 and 5.0 Hz)	
(e) $C_6H_5CHBrCH_2Br$	5.15(1H, dd, ca 6 and ca 9 Hz)	ABX (see Fig. 4.11)
(f) HOOCH(NH$_2$)CH$_2$COOH (in D$_2$O)	4.0(1H, dd, ca 6 and ca 8 Hz)	ABX (see Fig. 4.12)

The X portion of the ABX system is made up of six lines, symmetrical about the mid-point, the outer lines of which are of very low intensity so that it resembles the X portion of the AMX spectrum. An exact δ value can be read directly from the spectrum for H$_X$ but only approximate J_{AX} and J_{BX} values. The AB portion consists of the expected eight lines, the outside four of which are less intense than

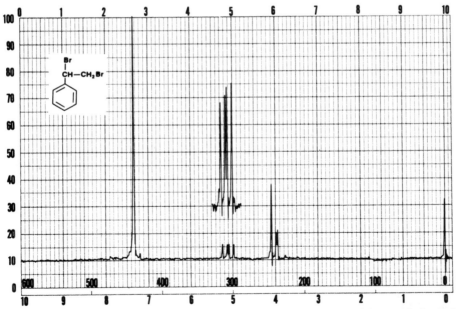

Figure 4.11 ABX spectrum: 1,2-dibromoethylbenzene. Reprinted with permission of Aldrich Chemical Co., Inc.

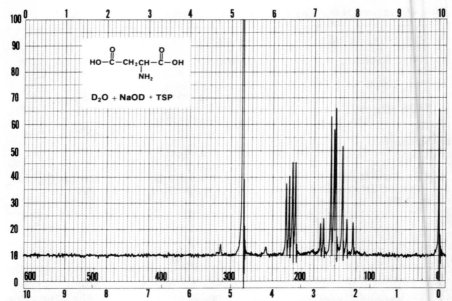

Figure 4.12 ABX spectrum: L-aspartic acid. TSP = 3-(trimethylsilyl)propionic acid-2,2,3,3-d_4 acid. As Na salt, water-soluble reference compound. Reprinted with permission of Aldrich Chemical Co., Inc.

the other four, leaving four strong lines corresponding to the two central lines of an AB spectrum but further split unequally by the third nucleus, H_X. Two of these strong lines sometimes coincide, as in the spectrum of 1,2-dibromo-1-phenyl-ethane (Fig. 4.11). Examples of moleculer fragments which give rise to these patterns are listed in Table 4.6 and illustrated again in Fig. 4.12.

4.7.3 The AX_3 system

As might be expected, this system consists of a doublet and a quartet as shown in Fig. 4.13. Examples of compounds or part structures which give rise to the system are given in Table 4.7.

The rarely encountered AB_3 spin system, containing up to fourteen lines, results when the $\Delta v/J$ ratio is small.

Table 4.7. Molecules that give rise to the AX_3 system

Molecule	δ(ppm)[J(Hz)]
(a) CH_3CHO	9.8(1H, q, 3.0 Hz); 2.2(3 H, d, 3.0 Hz)
(b) $CH_3CH=C(X)Y$	ca 6(1H, q, 4.5 Hz); ca 1.8(3H, d, 4.5 Hz)
(c) $CH_3CH(OEt)$	4.7(1H, q, 6.0 Hz); 1.2(3H, d, 6.0 Hz)

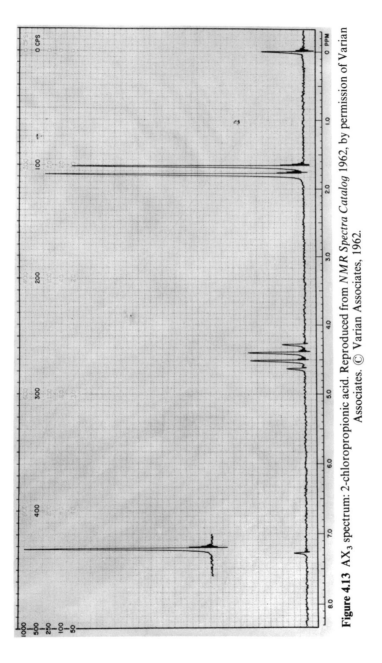

Figure 4.13 AX$_3$ spectrum: 2-chloropropionic acid. Reproduced from *NMR Spectra Catalog* 1962, by permission of Varian Associates. © Varian Associates, 1962.

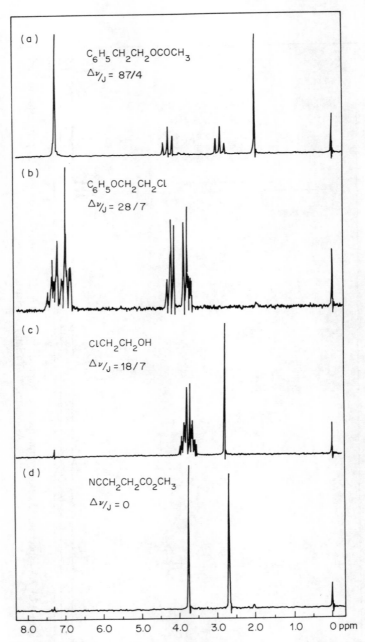

Figure 4.14 Change in the A_2X_2 spectrum with decrease in the $\Delta v/J$ ratio.

4.7.4 The A_2X_2 system

The characteristic two sets of triplets produced by this system are shown in Fig. 4.14a. As the $\Delta v/J$ ratio decreases over the range ca 4–0 the splitting pattern for the $-CH_2CH_2-$ system becomes progressively more complicated, until at $\Delta v/J = 1-2$ a maximum of 24 major lines may be observed as in Fig. 4.14c. Neither J nor δ values can be read directly from this spectrum but the A_2B_2 system, as it has now become, is recognizable by a 2:2 or a 4H integral and by the mirror-image relationship the A_2 portion bears to the B_2 portion. At smaller ratios, more closely spaced lines are seen, until ultimately at $\Delta v = 0$ a four-portion singlet (see Fig. 4.14b and e), is produced. Table 4.8 lists other structures containing this spin system.

Table 4.8. Molecules that give rise to the A_2X_2 system

Molecule	δ(ppm) [J(Hz)]	System
(a) $C_6H_5CH_2CH_2OCOR$	4.3 (2H, t, 7.0 Hz); 2.9 (2H, t, 7.0 Hz)	A_2X_2
(b) $HOCH_2CH_2CN$	3.9 (2H, t, 6.5 Hz); 2.6 (2H, t, 6.5 Hz)	A_2X_2
(c) $HOCH_2CH_2Cl$	ca 3.9 (2H, m); ca 3.75 (2H, m)	A_2B_2
(d) $NCCH_2CH_2COOR$	2.7 (4H, s)	A_4
(e) $X_2CHCH_2CHX_2$	(2H, t); (2H, t)	A_2X_2

Note. It has already been remarked (Section 4.7.2.4) that many compounds giving spectra with non-first-order splitting patterns at lower applied field strengths give first-order patterns at higher field strengths. This is a consequence of a change in the $\Delta v/J$ ratio and, since J is a constant, Δv must vary. This is indeed the case as although the chemical shift expressed in ppm from TMS is a constant, the number of hertz per part per million shift depends on the applied field strength (see Section 4.9.8). The spectrum of 2-chloroethanol (Fig. 14c at 60 MHz) appears at 400 MHz, where 1 ppm = 400 MHz, as in Fig. 14a, the $\Delta v/J$ ratios being 9/7 and $(9 \times 400/60)/7 = 60/7$, corresponding to A_2B_2 and A_2X_2 systems, respectively.

4.7.5 The A_2X_3 and AX_6 systems

These systems are most commonly represented by ethyl and isopropyl groups, as illustrated in Figs 4.15 and 4.16. Since methyl protons are usually well separated from methylene protons, as are methyl from methine protons, such combinations rarely produce higher order splitting patterns.

4.7.6 Aromatic compounds

The interpretation of signals in the aromatic region is always helped by reference to the chemical shift value (7.27 ppm) of the benzene protons. Shielding and deshielding effects are quoted in the following section from this reference position.

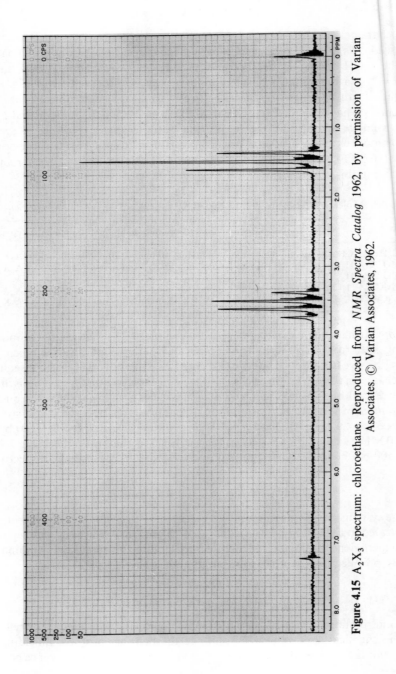

Figure 4.15 A_2X_3 spectrum: chloroethane. Reproduced from *NMR Spectra Catalog* 1962, by permission of Varian Associates. © Varian Associates, 1962.

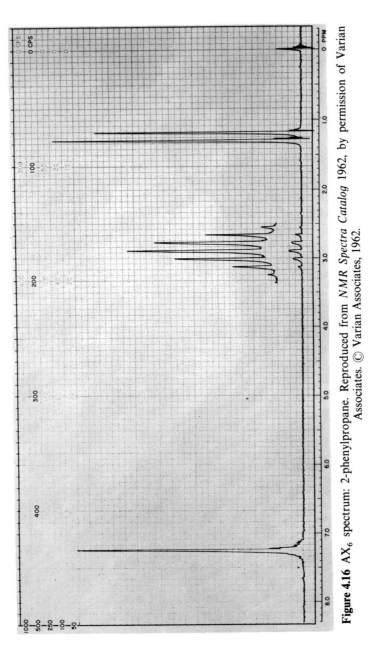

Figure 4.16 AX$_6$ spectrum: 2-phenylpropane. Reproduced from *NMR Spectra Catalog* 1962, by permission of Varian Associates. © Varian Associates, 1962.

4.7.6.1 Alkyl substitution

Alkyl groups cause little perturbation of the aromatic π system. In consequence, arylalkanes such as toluene, 1,4-dimethylbenzene (see Fig. 4.17) and 1,3,5-trimethylbenzene give rise to spectra consisting of two sharp singlets of appropriate intensity, one in the aromatic and the other in the alkane proton region. Even with less symmetrically substituted arylalkanes the above pattern is approximately adhered to. The aromatic protons experience only a slight upfield shift (0.1–0.3 ppm).

4.7.6.2 +M Substituents

The +M effect operating in aromatic nuclei due to substitution by groups such as amino, hydroxy and alkoxy increases the electron density at the *ortho* and *para* carbons, inducing shifts to higher field (shielding) of the attached protons. As a result, the spectrum of anisole shows a 3H:2H division between multiplets centred at 6.85 and 7.20 ppm, respectively. This effect is even more marked in the case of amines, as can be seen in Fig. 4.18a. Complete resolution of the aromatic protons can be achieved at 270 MHz (Fig. 4.18b).

4.7.6.3 −M Substituents

When the substituent is of the −M type such as carbaldehyde, nitro or amido, anisotropic and mesomeric effects combine to deshield the *ortho* protons by up to 1 ppm, whereas the *para* proton, affected only by the mesomeric effect, is deshielded by only *ca* 0.3 ppm. A 2:3 or a 2:1:2 integral sequence is therefore seen, again in keeping with the pattern of charges in the $C_6H_5X = Y$ (where Y is more electronegative) charge-separated canonical forms. This is illustrated in Fig. 4.19 by the spectrum of benzaldehyde.

4.7.6.4 Disubstituted benzenes

If the substituents are identical and *para* positioned, then a four-proton singlet is seen. If they are different, an AA'XX' system is often observed, as in the spectrum of 4-nitroanisole shown in Fig. 4.20a. When identical substituents are located *ortho* to each other, the AA'XX' generally consists of more lines but still has a symmetry plane (see Fig. 4.20b), a feature also of the related AA'BB' system. It should be noted that the appearance of the AA'XX' pattern can vary greatly; thus the furan spectrum, 7.42 and 6.37 ppm (2H, t, 3.0 Hz), results from a coincidence of J_{AX} and $J_{AX'}$. To explain the designation of the proton spin systems encountered in, for example, 4-nitroanisole and furan as AA'XX', we need to define magnetic equivalence more closely as nuclei having the same chemical shift and being coupled equally with all other nuclei in the molecule. In both of these cases the

NUCLEAR MAGNETIC RESONANCE SPECTROSCOPY

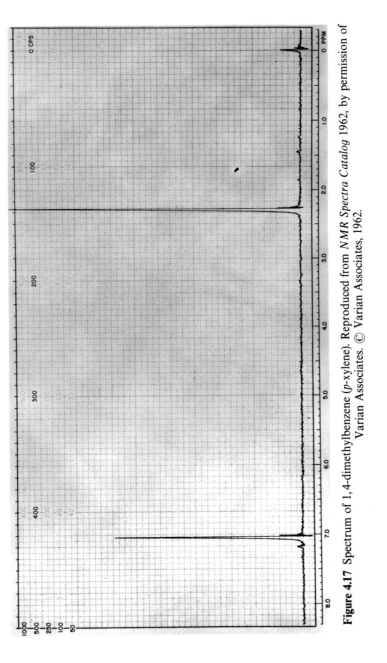

Figure 4.17 Spectrum of 1,4-dimethylbenzene (p-xylene). Reproduced from *NMR Spectra Catalog 1962*, by permission of Varian Associates. © Varian Associates, 1962.

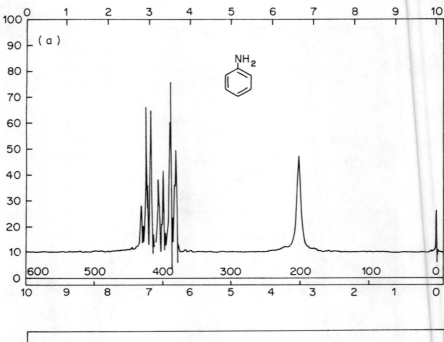
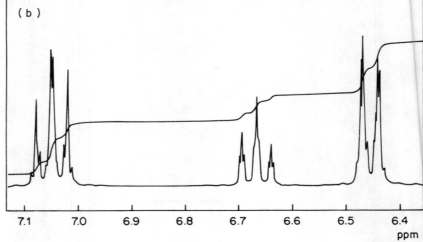

Figure 4.18 (a) 60 MHz spectrum and (b) 270 MHz spectrum of aniline aromatic region.

NUCLEAR MAGNETIC RESONANCE SPECTROSCOPY

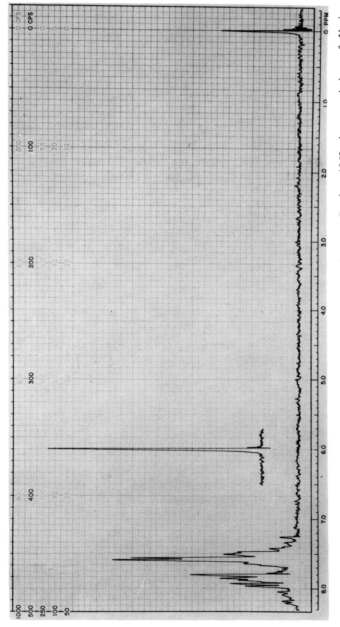

Figure 4.19 Spectrum of benzaldehyde. Reproduced from *NMR Spectra Catalog* 1962, by permission of Varian Associates. © Varian Associates, 1962.

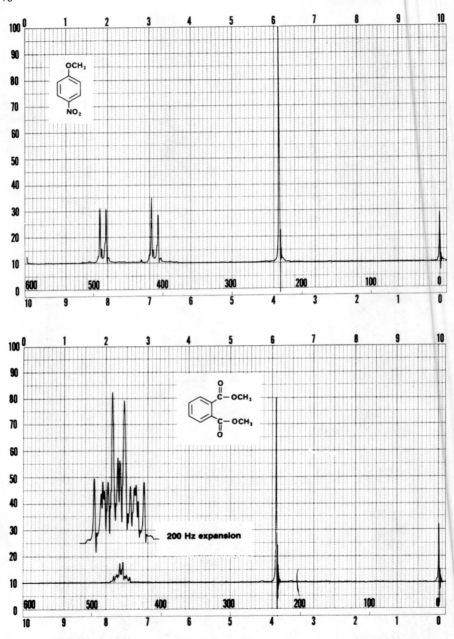

Figure 4.20 AA'XX' spectra: (a) *p*-nitroanisole; (b) dimethyl phthalate. Reprinted with permission of Aldrich Chemical Co., Inc.

high $\Delta v/J$ ratios justify the use of the letters A and X, and the protons assigned the same letters have identical chemical shifts. However, H_A is coupled differently to H_X than is $H_{A'}$ and likewise $H_{A'}$ is coupled differently to $H_{X'}$ than is H_A, and so on. This is not the case in the A_2X_2 examples cited in Section 4.7.4, where both the A nuclei are identically coupled to the X nuclei and *vice versa*. In fact, the AA'XX' system gives rise to 20 lines. It is simply the high-intensity bands which give the characteristic line patterns referred to in the discussion. Some of the other lines are observed as spikes around the major bands, some are overlapped by them and still others may be lost in the background noise.

4.7.6.5 Trisubstituted benzenes

Compounds of the type $C_6H_3X_2Y$ such as 2,6-dinitroaniline and 3,5-dinitrotoluene give rise to typical A_2X splitting patterns. The 1,2,4-trisubstitution pattern is frequently met and, if $J_{1,4}$ is significant, corresponds to the twelve-line AMX array described in Section 4.7.2.3. Most often, however, $J_{1,4} = 0$, so that only eight lines are apparent as in Fig. 4.21. Depending on the nature of substituents, the δ values of the sets of absorptions, two doublets and one doublet of doublets, allow a variety of arrangements of this basic pattern, two being shown in Fig. 4.21.

4.8 INTERPRETATION OF PROTON NMR SPECTRA

In real life, it is likely that you would have made the compound under examination from known starting compounds with a structural objective in mind. You would also have other spectral data to hand and possibly have developed some familiarity with the spectral properties of the class of compounds to which the unknown belongs. Hence this approach, i.e. the identification of a compound from just one type of spectrum and a molecular formula, is unrealistic, but it does discipline the student into wringing as much information as possible from the data provided.

Normally, organic chemists use their knowledge of chemistry to predict possible products from a reaction and then see which structure fits the spectral data best. There is always a danger in this that more than one structure may be consistent with *most* of the data *and* the chemistry. The first one which comes to mind, however, is often believed to be correct, the idea of alternative, possible structures being unconsciously excluded. It is therefore most unwise to ignore 'the odd inconsistency.' All the features of proton and carbon NMR spectroscopy should be explicable in terms of the proposed structure. The best practice for the beginner is to interpret spectra fully on the basis of the given, correct structure.

80　　　　　　　　　　　　　　　　　　　　　　　　　　　ORGANIC SPECTROSCOPY

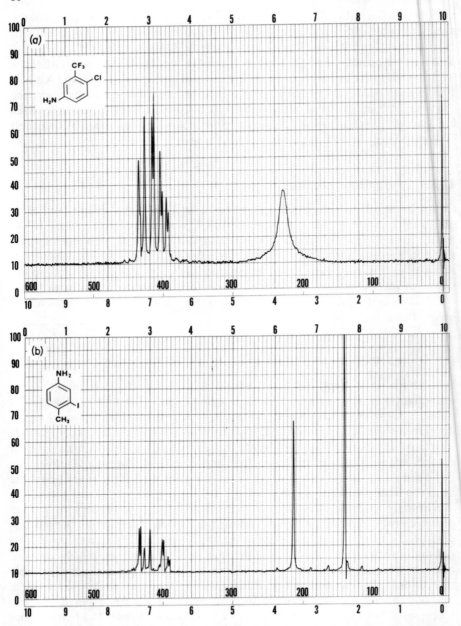

Figure 4.21 AMX spectra, $^5J_{AX} = 0$; (a) 4-chloro-2-trifluoromethylaniline; (b) 3-iodo-4-methylaniline. Reprinted with permission of Aldrich Chemical Co., Inc.

4.8.1 Example 1

Compound 1, $C_{18}H_{35}N$, gives the spectrum shown in Fig. 4.22.

(a) From the molecular formula, the double bond equivalent (DBE) (see Section 5.6) is 2.
(b) The integral ratio is 3:28:2:2, consistent with the molecular formula.
(c) $\delta\,0.9\,(3H, t, J \approx 6\,Hz)$ indicates a methyl group adjacent to a methylene group in an alkyl chain, i.e. CH_3CH_2—alkyl.
(d) $\delta\,2.35\,(2H, t, J \approx 7\,Hz)$ indicates a methylene group adjacent to another methylene group but also attached to a deshielding group, i.e. $-CH_2CH_2X$. The shielding effect causes a 1.1 ppm downfield shift.
(e) $\delta\,1.2-1.7\,(26H, 2H, 2H)$ indicates an alkyl chain of 15 carbons with one or two of the methylenes slightly deshielded, probably β and γ, to a $-I$ group.
(f) The whole suggests $CH_3(CH_2)_{15}CN$ or $CH_3(CH_2)_{15}NC$ as the structure of compound 1. This fits the DBE. The J values are consistent with alkyl chain ^3J (see 4.16.3) couplings. Use of Table 4.19 (Appendix 4.16.1.4) to calculate the δ value for $-CH_2CN$ would suggest that the former is the correct structure, although no incremental shift is provided for the isocyanide function to allow a comparison.

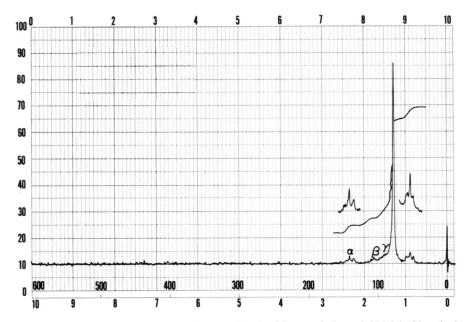

Figure 4.22 Spectrum of compound 1. Reprinted with permission of Aldrich Chemical Co., Inc.

4.8.2 Example 2

Compound 2, $C_9H_{11}BrO$, gives the spectrum shown in Fig. 4.23.

(a) DBE = 4.
(b) The integral ratio 5:2:2:2 is consistent with the molecular formula.
(c) δ 2.25 (2H, pentet, J = 7 Hz): the multiplicity indicates two possible part structures, i.e. CH_3CH_2CH- or $-CH_2CH_2CH_2-$. The former is ruled out as there is no triplet around 0.9 ppm.
(d) δ 3.35 and 4.05 (2H, t, J = 7 Hz) each pointing towards the high-field pentet. This is consistent with the second part structure shown above. Both methylenes are deshielded and so are attached to electronegative groups, i.e. $XCH_2CH_2CH_2Y$.
(e) δ 6.8–7.4 (5H, m) indicates a C_6H_5 unit, substituted by O or N since some of the hydrogens are shifted above the benzene δ value, hence C_6H_5O = X or Y.
(f) This leaves only Br in the molecular formula unaccounted for. Hence Br = X or Y and the structure of compound 2 is $C_6H_5OCH_2CH_2CH_2Br$.

Allocation of the individual methylene groups can be made by application of Table 4.19 in Appendix 4.16.1.4.

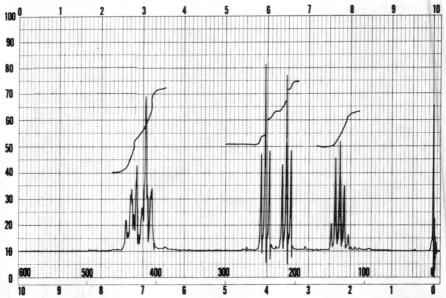

Figure 4.23 Spectrum of compound 2. Reprinted with permission of Aldrich Chemical Co., Inc.

4.8.3 Example 3

Compound 3, C_5H_9N, has the spectrum shown in Fig. 4.24. The singlet at 1.65 ppm is removed by D_2O.

NUCLEAR MAGNETIC RESONANCE SPECTROSCOPY

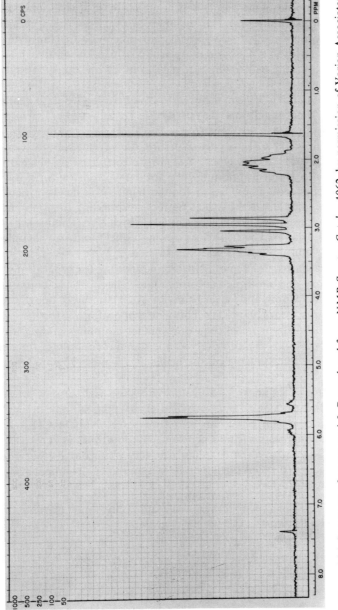

Figure 4.24 Spectrum of compound 3. Reproduced from *NMR Spectra Catalog* 1962, by permission of Varian Associates. © Varian Associates, 1962.

(a) DBE = 2.
(b) The integral shows a ratio from low to high field of 2:2:2:2:1. By comparing this with the molecular formula we know that these are the actual number of protons corresponding to each signal.
(c) The one proton singlet at 1.65 ppm is removable by D_2O. Hence it is likely to be one of the proton types listed in Appendix 4.16.2. The δ value indicates that it is probably attached to the nitrogen atom of an amine, which must be secondary.
(d) Since the signal at 2.95 ppm is a well resolved triplet, we shall deal with this next. The integral tells us that it is due to a —CH_2— group and its multiplicity that it is adjacent to $3 - 1 = 2$ equivalent protons ($n + 1$ rule), indicating a —CH_2CH_2— part structure. As it is not split further, the part structure is —CH_2CH_2X where X is fairly electronegative since it shifts the CH_2 group to about 3 ppm and does not carry a proton (or if it does, it is rapidly exchanged). The J value for the triplet (the distance between any consecutive lines, divided by the length corresponding to 1 ppm, multiplied by 60) is approximately 6 Hz [(2/19) × 60], which is reasonable for the ^3J in a normal alkyl chain. The triplet is pointing upfield and so is coupled to the two-proton multiplet at 2.1 ppm.
(e) δ 2.1 (2H, seven or eight lines, 21 Hz bandwidth, 3 Hz separations evident). The width of the multiplet accommodates the 2×6 Hz coupling to the methylene at 2.9 ppm and leaves room for other couplings, one of which from the spectrum appears to be 3 Hz. This would allow for one further coupling of $21 - 12 - 3 = 6$ Hz or more than one smaller couplings adding up to 6 Hz. The latter possibility is more likely as no other 6 Hz coupling is evident in the spectrum. The fine splitting on the peaks at 3.3 and 5.75 ppm seems to link them.
(f) δ 3.3 (2H, five or six lines). The δ value places this CH_2 also next to nitrogen, but it is further deshielded by approximately 0.4 ppm by some other structural feature. Hence we have the part structure —$CH_2NHCH_2CH_2$—.
(g) δ 5.7 (2H, m) suggests olefinic protons. The multiplet is reminiscent of an AB, $J \approx 12$ Hz, where the δ values are very close but further coupling is present. The two protons must then be in a similar environment and probably have a Z relationship.

If we put the pieces together, we find only one possible structure: —$CH_2NHCH_2CH_2$ and —CH=CH— gives

$$\begin{array}{c} H \\ | \\ C \\ H_2C \diagup \diagdown CH \\ H_2C \diagdown \diagup CH_2 \\ N \\ | \\ H \end{array}$$

As further confirmation of the structure, we can see that the methylene protons at 3.3 ppm have a signal width of about 10 Hz. This is acceptable in the part structure (Z)—$XCH_2CH{=}CHCH_2Y$ in view of the possible ^3J, ^4J, ^5J couplings (see Appendix 4.16.3.2) through the allyl system. The other allylic methylene protons should therefore have a bandwidth of $10 + 6 + 6 = 22$ Hz, which on inspection proves to be so.

4.9 BASIC CONCEPTS

4.9.1 Nuclear magnetism

The nuclei of certain atoms possess both mechanical spin and electric charge. As a consequence, they generate a magnetic field and can therefore be likened to bar magnets having characteristic properties of spin and magnetic moments. These properties are shared by both protons and, surprisingly at first sight, by neutrons. The latter may be regarded as undergoing reversible decay to protons and electrons so that the uneven distribution of charge within the spinning neutron gives rise to a magnetic field.

The degree of spin is defined by the nuclear spin quantum number, I, which increases in half integral values from zero, so that $I = 0, \frac{1}{2}, 1, 1\frac{1}{2}, 2$, etc. All nuclei with an odd mass number are magnetic, as are even mass numbered nuclei with odd numbers of protons and neutrons. It follows that ^1H, ^{13}C, ^{15}N, ^{17}O, ^{19}F and ^{31}P are magnetic and may generate NMR spectra, whereas nuclei such as $^{12}_{6}C$, $^{16}_{8}O$ and $^{32}_{16}S$ cannot. The nuclei which have so far proved most amenable to the generation of NMR spectra, viz. ^1H, ^{13}C, ^{19}F and ^{31}P, are those having $I = 1/2$. The magnitude of I cannot be predicted (see Table 4.9).

4.9.2 Nuclei in a magnetic field

If a magnetic nucleus is placed in a magnetic field, unlike the classical bar magnet which will align itself with the field, it can take up any one of $2I + 1$ orientations. These orientations are the result of the different interactions determined by quantum mechanical calculations between the magnetic nucleus and the applied field. When $I = 1/2$ there are two nuclear magnetic energy levels, $m_I = \pm 1/2$, describing the two possible orientations, viz. $+1/2$ or $-1/2$ with respect to the applied field. If the spin is aligned (parallel) to the applied magnetic field, the nucleus is regarded as having a lower energy, $m_I = 1/2$, sometimes represented by $\uparrow$ or α. The $m = -1/2$ ($\downarrow$ or β) level is then said to be spin opposed or antiparallel to the applied field. In the absence of an applied field the spin states are degenerate. The difference in energy between the two spin states is given by

$$\Delta E = \gamma B_0 h / 2\pi$$

where γ is the gyromagnetic ratio having a characteristic value for each nucleus,

B_0 is the applied field strength (in Tesla) and h is Planck's constant. The gyromagnetic ratio is the ratio of magnetic moment to angular momentum; the latter is quantized in units of $h/2\pi$. It follows that γ has the units $s^{-1}T^{-1}$.

4.9.3 Frequency required to induce transitions

The difference in energy, ΔE, between the spin states is a function of the applied magnetic field and also of the gyromagnetic ratio, γ, of the nucleus, where

$$\gamma = \frac{\text{magnetic moment}}{\text{angular momentum}}$$

and angular momentum is quantized in units of $h/2\pi$. Hence

$$\Delta E = f(B_0\gamma) = h\nu$$
$$\Delta E = \gamma B_0 h/2\pi = h\nu$$

If we solve for the frequency of absorption necessary to induce a transition, then

$$\nu = \gamma B_0/2\pi$$

For the proton in an applied field of 1.41 T, the required frequency is obtained from

$$\nu = \frac{2.6738 \times 10^8 \, s^{-1} T^{-1}}{2 \times 3.141} \times 1.41 \, T = 60 \times 10^6 \, s^{-1} \text{ or } 60 \text{ MHz}$$

Hence a radio-frequency of 60 MHz is necessary.

4.9.4 Energy between spin states

The energy required to bring about the transition of one proton is $\Delta E = h\nu = 6.63 \times 10^{-34} \times 60 \times 10^6 = 4 \times 10^{-26} J$, and per mole is $4 \times 10^{-26} \times$ Avogadro's number $(6.02 \times 10^{23}) = 2.4 \times 10^{-2} J$.

4.9.5 Population of spin states

Knowing the difference in energy between any two spin states in equilibrium, it is possible to calculate the populations of the spin states using the Boltzmann distribution. For the proton we have seen that $\Delta E = 0.024 \, J \, ml^{-1}$. Hence

$$N_\alpha/N_\beta = \exp(-0.024/RT) = \exp[-0.024/(1.98 \times 4.2 \times 300)]$$
$$= \exp(-9.62 \times 10^{-6})$$
$$N_\alpha/N_\beta = 1.00001$$

This indicates that for every one million and ten nuclei in the α spin state there are one million in the β spin state. Since the radio-frequency employed to bring about transition between the two spin states does so with equal probability, then the maintenance of a net absorption depends on the maintenance of the population difference. However the applied radio-frequency tends to equalize populations since there are more transitions from the more populated lower to the higher energy level. Thus continuous irradiation tends to an equalization of populations of the spin states and a corresponding diminution in absorption signal strength, ultimately to zero. This phenomenon is termed saturation. However, opposing this tendency are a number of relaxation processes (see Section 4.9.6) in which transitions, often called 'spin flips,' occur from higher to lower energy without the emission of radio-frequency energy.

4.9.6 Relaxation processes

As stated, the relaxation processes which restore the Boltzmann population distribution do so by radio-frequency radiationless processes. The most important process for losing magnetic energy is called spin–lattice relaxation and requires interaction with local fluctuating magnetic fields of correct magnitude and orientation. These local fields are associated mainly with solvent molecules, the more polar, e.g. $CHCl_3$ and CF_3COOH being more efficient relaxing agents. The magnetic energy is given out as heat energy.

A second spin–lattice relaxation process is due to dipole–dipole interactions between nearby nuclei. This is particularly important in ^{13}C NMR spectroscopy where the ^{13}C nucleus is relaxed by attached hydrogens in the $^{13}C-^{1}H$ system.

A third spin–lattice interaction is that between a magnetic nucleus and a paramagnetic species. When an odd-electron species is present in the sample, its effect usually dominates over the other relaxation, processes to such an extent that it often causes line broadening (see Section 4.9.7).

A fourth such process is that of spin–spin relaxation, by which two nuclei swap spins. This does not affect the Boltzmann distribution but reduces the lifetime of a given spin state.

4.9.7 Line broadening

The sharpness of an absorption line is dependent on the homogeneity of the applied field. However, this need not concern us since very stable homogeneous fields are now readily attainable. A more important broadening effect from the spectroscopist's viewpoint is that due to uncertainty. If the lifetime of a spin state is very short there will be uncertainty as to the energy associated with it; in other words, there will be a spread of energies. Heisenberg's equation can be expressed in this connection as follows:

$$\Delta E, \Delta t = h/2\pi$$

Since $E = h\nu$,

$$\Delta t = \frac{1}{2\pi \Delta \nu}$$

For a sharp line of say 1 Hz, $\Delta t = 1/(2\pi \times 1) = 0.16$ s. This shows that a sharp absorption line, ca 1 Hz, can be obtained from a spin state having a lifetime of about 0.16 s and further illustrates the resolving power of modern spectrometers as often much better than 1 in 60×10^6.

For a shorter lifetime of say 0.01 s, $\Delta v = 1/(2\pi \times 0.01) = 16$ Hz. This would give rise to an unacceptably broad signal.

In summary, if relaxation is too slow the nuclei becomes saturated and the absorption signal dippears; if the relaxation is too fast line broadening will occur (Δt and Δv are inversely proportional) and again the signal may disappear into the spectral baseline.

4.9.8 Chemical shifts

It has been shown (Section 4.9.3) that for a field strength of 1.41 T, spin flip and hence absorption of energy can be induced by irradiation of the sample by a 60 MHz radio-frequency source. This is a theoretical value for the unreal circumstance of an isolated proton. If all protons absorbed under precisely the same conditions then there would be no differentiation between different kinds of protons. In reality, each type of proton resonates under slightly different conditions because each type of proton finds itself in a slightly different magnetic field. That is the applied magnetic field reduced by local induced magnetic fields which are set up by and opposed to the applied field. This reduction may be quantified as the shielding constant, σ, in the equation

$$v = \frac{\gamma B_0 (1 - \sigma)}{2\pi}$$

No table of shielding constants based on this equation finds general use, since it is more convenient to define differences in absorption frequency by reference not to a bare proton but most often as a chemical shift downfield from the TMS absorption frequency (^{1}H or ^{13}C as the case may be) which is arbitarily placed at zero. From the above equation, it is evident that the resonance frequency is dependent on the applied magnetic field, so that the chemical shift parameter δ is independent of the field strength at which various spectrometers operate; δ is defined as

$$\delta = \frac{v(\text{sample}) - v(\text{reference})}{\text{working frequency}} \times 10^6 \text{ ppm}$$

where v(reference) is usually zero and v(sample) is quoted in cycles per second downfield from tetramethylsilane. If the methyl protons are separated by 228 Hz from those of TMS using a spectrometer with an applied field of 1.41 T and a frequency of 60 MHz, then the chemical shift of the methyl protons is

$$\delta(\text{CH}_3\text{O}) = \frac{228 - 0 \times 10^6}{60 \times 10^6} = 3.8 \text{ ppm}$$

In practice, the value is read directly from the chart paper calibrated in parts per million shift from TMS.

4.9.9 Spin–spin coupling. *J* the coupling constant

For an AX system (see Section 4.7.1), two doublets are observed. These arise as shown in Fig. 4.25, where for simplicity only nucleus A is considered.

(i) In the absence of an applied field the spin states are degenerate.
(ii) In the presence of the applied field B_0, nuclei such as ^{13}C, ^1H or ^{19}F with $I = 1/2$ can take up either one of $2I + 1 = 2$ orientations.
(iii) The applied field is modified in the region of A by the local field set up opposing it and further modified by the fields due to the difference spin states of nearby nuclei, which in this case will be $+\frac{1}{2}$ or $-\frac{1}{2}$, i.e. there will be two possible resultant local fields.
(iv) We can now imagine coupling to take place between the nuclei. The antiparallel arrangements (↑↓) or (↓↑) are stablizing and the parallel arrangements (↑↑) or (↓↓) are destabilizing. The two transitions now have different values.*

The two transitions can be drawn in the form of the observed splitting pattern (Fig. 4.26).

By the same considerations, the X portion of the AX spectrum would also appear as a doublet. It follows from this that the ethyl group, A_2X_3 system,

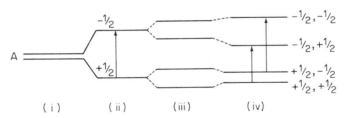

Figure 4.25

Figure 4.26

*The transitions are governed by selection rules such that only transitions $\Delta m_I = \pm 1$ are allowed, that is, $+\frac{1}{2}, -\frac{1}{2} \rightarrow -\frac{1}{2}, -\frac{1}{2}$ or $+\frac{1}{2}, +\frac{1}{2} \rightarrow -\frac{1}{2}, +\frac{1}{2}$.

consists of a triplet for the X_3 portion, since A_2 has the following four possible spin combinations, two of which are degenerate:

$$-\tfrac{1}{2}-\tfrac{1}{2}$$
$$-\tfrac{1}{2}+\tfrac{1}{2}, +\tfrac{1}{2}-\tfrac{1}{2}$$
$$+\tfrac{1}{2}+\tfrac{1}{2}$$

The A_2 portion appears as a quartet since it senses eight different spin combinations with two sets of three being degenerate:

$$-\tfrac{1}{2}-\tfrac{1}{2}-\tfrac{1}{2}$$
$$-\tfrac{1}{2}-\tfrac{1}{2}+\tfrac{1}{2}, -\tfrac{1}{2}+\tfrac{1}{2}-\tfrac{1}{2}, +\tfrac{1}{2}-\tfrac{1}{2}-\tfrac{1}{2}$$
$$-\tfrac{1}{2}+\tfrac{1}{2}+\tfrac{1}{2}, +\tfrac{1}{2}-\tfrac{1}{2}+\tfrac{1}{2}, +\tfrac{1}{2}+\tfrac{1}{2}-\tfrac{1}{2}$$
$$+\tfrac{1}{2}+\tfrac{1}{2}+\tfrac{1}{2}$$

Note that these spin distributions correspond to the intensity ratios of the A_2X_3 spin system, i.e. $1:2:1$ and $1:3:3:1$. It has already been stated that intensity ratios can be predicted from Pascal's triangle (see Section 4.5). They may also be predicted by using the coefficients of the binomial expansion $(a+b)^n$. For example, when $n=3, (a+b)^3 = a^3 + 3a^2b + 3b^2a + b^3 \equiv 1:3:3:1$. Quartets of this appearance are observed in the ^{13}C NMR spectra of the $^{13}CH_3$ and $^{13}CF_3$ groups. Only one eighth of the ^{13}C nuclei at any one time 'see' a given one of the eight possible spin combinations, but only four spectral lines result owing to the degeneracies referred to.

In the case of the methyl group the $^{13}C-H$ coupling constant, J, is 125 Hz and the $^{13}C-^{19}F$ coupling of the trifluoromethyl group is 280 Hz. Each of these is very large compared with the J_{H-H} coupling of ca 7.5 Hz in the ethyl group.

J values are extremely useful (see Appendix 4.16.3) in recognizing stereochemical relationships between protons, hybridization of carbons to which protons are attached and, in more complicated cases, tracing sequences of protons around a molecule by means of shared J values.

4.10 INSTRUMENTATION

4.10.1 Components

The following components are needed for a continuous-wave NMR spectrometer:

1. a stable magnet giving a homogeneous field between the poles;
2. an oscillator to provide stable radio-frequencies of suitable wavelength to induce nuclear spin transitions;

3. a receiver/amplifier to detect the absorption of energy;
4. a system to record, store, process and print out data.

The sample is placed in a glass tube and, in order to average any inhomogeneities in the field perpendicular to the direction of spinning, spun in the gap between the magnetic poles shown in Fig. 4.27. For a spectrometer operating at 1.41 T, energy is fed into the coil around the sample at 60 MHz for which frequency the detector is tuned. The radio-frequency source is kept constant and the magnetic field is swept by passing an increasing small current through the sweep coils by means of the sweep generator. Initially the radio-frequency detection system is balanced so that an even baseline appears until a nucleus comes into resonance. At this point the system is thrown out of balance and a resonance signal is produced. The continuous wave oscillator frequency cited above is for the detection of protons. Each type of nucleus requires a different frequency to excite transitions for a given narrow field-strength range. This is illustrated in Table 4.9.

Continuous-wave (CW) spectroscopy gives excellent results for the more sensitive nuclei, i.e. those of high abundance and large gyromagnetic ratio such as ^{1}H, ^{19}F and ^{31}P, but even using spectral accumulation techniques (see below) few other nuclei can be efficiently observed using CW spectroscopy.

The ^{13}C nucleus has a smaller gyromagnetic ratio than the proton. Consequently, for the same applied magnetic field there is a smaller energy difference between its spin states, and so a smaller population difference at equilibrium. As a result, the ^{13}C nucleus is inherently 0.016 times as sensitive to detection by NMR as the proton. If the low natural abundance of the ^{13}C nucleus is taken into

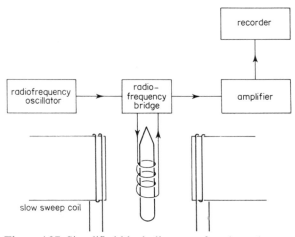

Figure 4.27 Simplified block diagram of nuclear absorption apparatus. Reproduced with permission from S. F. Dyke, A. J. Floyd, M. Sainsbury and R. S. Theobald, *Organic Spectroscopy*, 2nd ed., Longman, London, 1978.

Table 4.9. Nuclear properties of selected atoms

Nucleus*	Spin quantum number, I	Gyro-magnetic ratio, γ	R.f. (MHz)	Natural abundance (%)	Relative sensitivity
^{1}H	$\frac{1}{2}$	2.674	60	99.98	1
D	1	0.410	9.2	0.015	0.01
^{12}C	0			98.93	
^{13}C	$\frac{1}{2}$	0.672	15.1	1.07	0.016
^{14}N	1	0.193	4.3	99.63	0.001
^{16}O	0	0		99.76	
^{19}F	$\frac{1}{2}$	2.516	56.4	100	0.83
^{31}P	$\frac{1}{2}$	1.083	24.3	100	0.066
^{32}S	0			95.02	
^{29}Si	$\frac{1}{2}$	0.513	11.9	4.70	0.79

*Bromine and chlorine each have two isotopes with $I = \frac{3}{2}$.

account, we can see that the relative sensitivity of ^{13}C to ^{1}H in natural abundance is $0.016 \times 0.011 = 1.76 \times 10^{-4}:1$. In other words, the carbon nucleus is 5700 times less sensitive than the proton.

Instrumentation has now developed to the stage where natural abundance ^{13}C NMR spectra can be obtained very rapidly on 20 mg samples. The more obvious ways of increasing signal intensity are the use of higher applied field strengths (since $N_\alpha/N_\beta \propto B_0$) and repeated scan techniques whereby successive scans are accumulated and random noise tends to cancel. The signal enhancement is proportional to $\sqrt{\bar{n}}$ scans, so that 64 scans give an 8-fold enhancement, but a further 4032 scans (4096 total) are needed to obtain the next 8-fold (64-fold total) enhancement. However the breakthrough into the observation of the less sensitive nuclei has come about with the introduction of pulsed Fourier transform (PFT) spectrometers.

4.10.2 Pulsed Fourier transform spectroscopy

In this technique, the sample, as the name implies, is not subjected to continuous irradiation but to short, powerful pulses at set intervals in a constant magnetic field. These pulses are of only *ca* 50 ps duration and, by Heisenberg's uncertainty principle, we can see that a wide range of pulse frequencies is generated, bringing simultaneously all the nuclei into resonance. If the pulse length Δt is 50 ps, then since $h\Delta v, \Delta t \approx h$ and $\Delta v = 1/50 \times 10^{-6} = 20\,000$ Hz, we can see that even using a 500 MHz instrument, in theory the maximum likely frequency range of 10 000 Hz (20 ppm × 500 Hz) is covered. During the pulse, therefore, energy is absorbed as all the transitions are excited simultaneously. When it is switched off, the pulse-induced magnetism decays as the nuclei relax and the normal thermal Boltzmann distribution is restored. This is termed free induction decay (FID) and consists of

NUCLEAR MAGNETIC RESONANCE SPECTROSCOPY

a large number of decaying sine waves, each corresponding to the resonance frequency of a given nucleus or a given set of equivalent nuclei. This tangled skein is unravelled by a dedicated computer using Fourier analysis, which transforms the complex decay signal into the familiar intensity of absorption versus chemical shift plot obtained by the continuous wave method.

Thus a pulse takes microseconds and the FID and its collection a few seconds. The sequence is repeated many tens of times faster than one spectrum sweep by the CW method.

4.11 ^{13}C NMR SPECTROSCOPY

Before discussion of the interpretation of ^{13}C NMR spectra, the following points should be borne in mind.

1. In order to simplify the ^{13}C NMR spectrum, and for the reasons of extra sensitivity described below, the spectrum is usually obtained in the first instance with broad-band decoupling of protons. By this technique the whole range of protons is subjected to continuous, strong irradiation which shortens their nuclear spin lifetimes (see Section 4.2.1) so that the ^{13}C nuclei are decoupled from the neighbouring hydrogen nuclei and hence each type of carbon atom appears as a singlet absorption.
2. Couplings between adjacent ^{13}C nuclei are not observed in natural abundance spectra owing to the low probability of such a juxtaposition.
3. Off-resonance proton decoupling and some other techniques allow simplification of ^{13}C NMR spectra without losing all of the structural information provided by the multiplicity of the ^{13}C resonances. In Fig. 4.28 the broad-

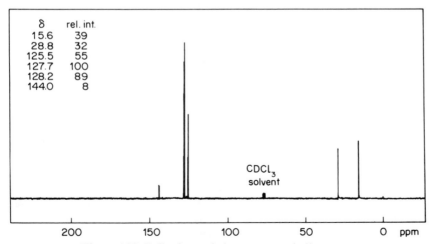

Figure 4.28 Fully decoupled spectrum: ethylbenzene.

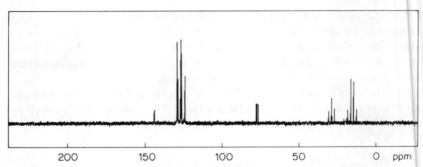

Figure 4.29 Off-resonance partially decoupled spectrum: ethylbenzene.

band or full proton-decoupled spectrum of ethylbenzene is reproduced and in Fig. 4.29 the off-resonance of partially proton-decoupled spectrum of ethylbenzene is shown to illustrate these two types of data presentation. Broad-band proton decoupling, which is used to decouple the ^{13}C nuclei from neighbouring protons, enhances the signal intensity by converting ^{13}C multiplets into singlets. Thus the ^{13}CH$_2$ group, where the carbon resonance would in the absence of decoupling appear as a triplet in 1:2:1 ratio, should become a singlet of relative intensity 4. In fact, the relative intensity is often of the order of 8 owing to the effect of the broad-band decoupling, which not only causes the multiplet to collapse to a singlet, but also affects the Boltzmann distribution of spin states to give a more favourable population distribution. This is called the nuclear Overhauser effect, and we shall return to this subject later (see Section 4.12.6).

The off-resonance spectrum indicates immediately for each carbon type whether it carries three, two, one or no hydrogen atoms according to the appearance (multiplicity) of each carbon resonance being a quartet, a triplet, a doublet or a singlet, respectively. A clearer way of presenting this information is, after processing the FT data, to display all carbons as singlets in two separate spectra. The first, termed 135° DEPT, shows all methyl and methine protons above the baseline (Fig. 4.30a) and all methylenes below, and the second, 90° DEPT (Fig. 4.30b), reveals the methine protons only. Quaternary protons are absent from these spectra and are detected by subtraction from the normal broad-band spectrum. In this way confusion arising from the overlap of multiplets due to carbon nuclei of similar chemical shifts is avoided. The explanation of the terms 90° and 135° DEPT is beyond the scope of this book.

4. In the PFT NMR technique, insufficient time is allowed for all ^{13}C nuclei to relax between pulses. The method is therefore not quantitative as is usually the case with ^{1}H NMR spectroscopy and so integral traces are not normally run. A semi-quantitative relationship does exist, however, between like carbon atoms, hence the approximately 2:2:1 ratio of intensities due to the aromatic

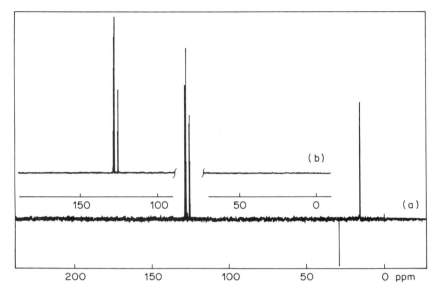

Figure 4.30 (a) 135° DEPT; (b) 90° DEPT: ethylbenzene.

protons in Fig. 4.28. Quantitative relationships between resonances can be achieved by the addition of Cr(acac)$_3$ to the sample solution. The paramagnetic chromium increases relaxion rates without producing too serious line broadening. Chemists are often reluctant to use this reagent, however, as the recovery of the sample is not always straightforward.

4.11.1 Interpretation of ^{13}C NMR spectra

4.11.1.1 Example 1

Compound 1, $C_6H_{10}O_3$ (DBE = 2), gives the spectra shown in Fig. 4.31a and b. The first spectrum shows six carbon types in addition to TMS at 0.0 ppm and the triplet due to CDCl$_3$ centred at 77 ppm, so there are no coincident peaks. The carbons between 20 and 40 ppm are alkanic and that at 52 ppm is alkanic but probably attached to an electronegative atom. The low-intensity carbons at about 170 and 200 ppm fall into the ester/amide and the ketone/aldehyde regions of the spectrum, respectively. This tallies with the DBE.

The spectrum in Fig. 4.31b indicates, from high to low field, the molecular fragments CH_2, CH_3, CH_2, CH_3, C and C. The compound is therefore a ketone rather than an aldehyde and the molecular formula requires the higher field quaternary carbon to be due to an ester rather than an amide. Note that the only low-field alkane carbon is due to a methyl group which must be present as a

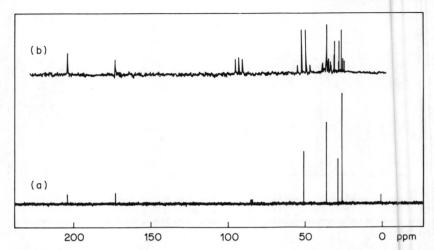

Figure 4.31 Compound 1: (a) fully decoupled spectrum; (b) off-resonance decoupled spectrum.

methyl ester. Note also the potentially confusing overlap of the two high-field multiplets.

We are left with the fragments C, COOMe, CH_2, CH_2 and CH_3. The three ways in which these can be put together are as follows: (a) $CH_3CH_2CH_2COCOOMe$; (b) $CH_3CH_2COCH_2COOMe$; and (c) $CH_3COCH_2CH_2COOMe$. If we use Table 4.10 (see later) to calculate the positions of the high-field methyl groups in these three structures, we find (a) $\delta = -2.3 + 9.1 + 9.4 - 3.0 = 13.2$; (b) $\delta = -2.3 + 9.1 + 3.0 - 2.5 = 7.3$; and (c) $\delta = -2.3 + 22.5 + 9.4 - 2.5 = 27.1$ (observed 29.6 ppm). Structure (c) therefore must be the correct one.

4.11.1.2 Example 2

Compound 2, $C_9H_{10}O$ (DBE = 5), gives the spectra shown in Fig. 4.32a and b. From Fig. 4.32a we can count seven carbon types in the molecule. This does not fit with the molecular formula unless two coincidences are assumed. Each of the intense absorptions might possibly correspond to two carbons in identical environments. The δ values suggest the presence of alkane carbons, 15–30 ppm, aromatic or olefinic carbons, 128–155 ppm, and a ketonic or aldehydic carbon, *ca* 200 ppm.

Scanning the spectrum in Fig. 32b from high to low field, we find that the molecule contains the molecular fragments CH_3, CH_2, CH, CH, C, C and CH. The low-field carbon is therefore aldehydic. Assuming that the two methine resonances are due to two carbons each, which is reasonable since they are more

NUCLEAR MAGNETIC RESONANCE SPECTROSCOPY

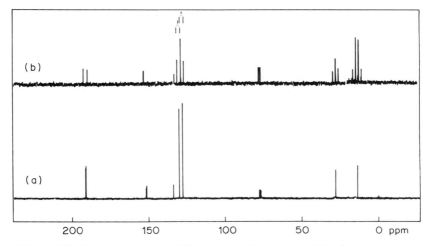

Figure 4.32 Compound 2: (a) fully decoupled spectrum; (b) off-resonance decoupled spectrum.

than twice the intensity of any other carbon present and further this would give a total of six carbons in the aromatic region, we now have a part structure (a).

Since two of the aromatic carbons are of such low intensity, disubstitution is indicated and the appearance of only four aromatic-type carbons necessitates the symmetrical substitution patterns shown in (b) and (c). Using Table 4.10 and taking only one representative carbon atom, we calculate the positions of the methylene groups in the two compounds: (c) $\delta = -2.3 + 29.9 + 22.1 = 49.7$; (b) $\delta = -2.3 + 9.1 + 22.1 = 28.9$ (observed 29.2 ppm). The structure of compound 2 is therefore (b).

4.11.2 Correlation tables

To identify the above compounds we have used correlation tables, the use of which is illustrated more fully later. The more important chemical shift values of protons are fairly readily memorized. As a consequence, students unfortunately tend to ignore the proton correlation tables by which δ values can be readily assessed to $ca \pm 0.3$ ppm, and thus miss the opportunity to make finer judgements between structural alternatives than can be made by reliance upon memory. The greater dispersion, 200 ppm, characteristic of ^{13}C NMR allows a greater precision in the assignment of chemical environments to ^{13}C nuclei than is the case with proton nuclei. To do this one cannot avoid the use of shift correlation tables. Those provided in this book are calculated and approximated from shift data aquired from simple, monofunctional compounds and so the results obtained by using them for more complicated, polyfunctional molecules should be treated with a degree of healthy scepticism. However, they will give great help for little effort, being very easily handled on only a short aquaintance.

4.11.2.1 Sequence of shifts

A brief inspection of Tables 4.10–4.15 confirms that the same sequence of shifts from high to low field, viz. alkanes–alkynes–alkyl chlorides, alcohols and fluorides–alkenes–aromatics–aldehydes, holds for both carbons and protons. Anisotropic effects operating over long distances do not appear to be important in carbon spectroscopy.

The δ values of alkane carbons are very sensitive to both crowding and inductive effects, so in calculating δ values for branched molecules a crowding term is introduced. Carbons carrying iodine or bromine appear at much higher field than those carrying chlorine or fluorine owing to shielding by the larger, more polarizable halogens and their lower electro-negativities.

4.11.2.2 Alkanes and substituted alkanes

The δ value of a carbon atom ^iC can be calculated as the sum of a series of constants so that

$$\delta = -2.3 + A + B$$

where A is the sum of the increments allowed for various substituents depending on their positions as α, β or γ to the ^{13}C atom in question, B is the sum of the branching corrections and -2.3 is the δ value for methane. Increments for A and B are given in Tables 4.10 and 4.11, respectively.

The functional groups shown in Table 4.11 may, for the purpose of selecting the steric correction increment, be regarded as corresponding to the types of carbon atom cited. In selecting the increment for the B term, for amines and ethers, regard the heteroatom as a carbon atom. O- and N-alkyl groups in esters and amides are counted as γ-substituents in calculating the value of A for ^iCCXR. There is no β increment for amide, ester, alkene, alkyne, phenyl, etc., as it is incorporated in the α increment.

Table 4.10. Increments for A, the shielding term

Substituent	Increment			Substituent	Increment		
	α	β	γ		α	β	γ
—C(sp³)	9.1	9.4	−2.5	—CHO	29.9	−0.6	−2.7
—C≡C—	4.4	5.6	−3.4	—CO—	22.5	3.0	−3.0
—C=C—	19.5	6.9	−2.1	—COOH	20.1	2.0	−2.0
—Ph	22.1	9.3	−2.6	—COO⁻	24.5	3.5	−2.5
—CN	3.1	2.4	−3.3	—COOR	22.6	2.0	−2.8
—O—	49.0	10.1	−6.2	—CON<	22.0	2.6	−3.2
—O—CO—	56.5	6.5	−6.0	—Cl	31.0	10.0	−5.1
—N<	28.3	11.3	−5.1	—F	70.1	7.8	−6.8
—N⁺<	30.7	5.4	−7.2	—Br	18.9	11.0	−3.8
—NO₂	61.6	3.1	−4.6	—I	−7.2	10.9	−1.5
—epoxide	21.4	2.8	−2.5	—S—	10.6	11.4	−3.6

Table 4.11. Increments for B, the branching or steric correction term

¹³C atom observed	Nature of the α-substituents			
	Primary	Secondary	Tertiary	Quaternary
Primary	—	—	−1.1	−3.4
Secondary	—	—	−2.5	−7.2
Tertiary	—	−3.7	−9.5	−15.0
Quaternary	−1.5	−8.4	−15.0	−25.0

Carbon equivalent	Functional group
Primary	$CO_2H, CO_2R, NO_2, OH, SH, NH_2$
Secondary	$C_6H_5, CHO, CONH_2, CH_2X,$* COR, OR, NHR, SR

*$X = OH, NH_2, SH$ or halogen.

The above points are illustrated by the following calculations:

(a) $CH_3CH_2{}^iCH(OH)CH_3$:

$$\delta^i = -2.3 + A + B$$
$$= -2.3 + (2\alpha + \alpha + \beta) + B$$
$$= -2.3 + (9.1 \times 2 + 49.0 + 9.4) + (-3.7)$$
$$= 70.6 \text{ (observed 68.8).}$$

(b) $CH_3{}^iCH_2CH(OH)CH_3$:

$$\delta^i = -2.3 + (9.1 \times 2 + 9.4 + 10.1) + (-2.5)$$
$$= 32.9 \text{ (observed 32.3).}$$

(c) $(CH_2OH)_3{}^iCNO_2$:

$$\delta^i = -2.3 + (9.1 \times 3 + 61.6 + 10.1 \times 3) + (-8.4 \times 3 - 1.5)$$
$$= 90.2 \text{ (observed 94.9)}.$$

(d) $(CH_3)_3{}^iCCOCH_3$:

$$\delta^i = -2.3 + (9.1 \times 3 + 22.5 + 9.4 + 0.0) + (-1.5 \times 3 - 8.4)$$
$$= 44.0 \text{ (observed 44.3)}.$$

(e) $CH_3\overset{O}{\overset{\|}{C}}CH_2{}^iCH_2\overset{O}{\overset{\|}{C}}OCH_2CH_3$:

$$\delta^i = -2.3 + (9.1 + 22.6 + 3.0 - 2.5 - 2.5)$$
$$= 27.4 \text{ (observed 27.8)}.$$

(f) $CH_3{}^iCHCH_2CH_3$:
   $\quad\quad\;\;|$
   $\quad\;\;OCH_3$

$$\delta^i = -2.3 + (9.1 \times 2 + 49.0 + 9.4 \times 2) + (-3.7 \times 2)$$
$$= 76.3 \text{ (observed 77.0)}.$$

4.11.2.3 Alkenes and alkene derivatives

The δ value of a carbon atom ^iC can be calculated as the sum of a series of constants such that

$$\delta^i = 122.8 + A$$

where A is the sum of the increments allowed for substituents on and β to ^iC and 122.8 is the observed δ value for ethene.

Increments for A for alkenes are given in Table 4.12.

4.11.2.3.1 Examples

(a) (E)—$CH_3CH={}^iCHCH_2CH_3$:

$$\delta^i = 122.8 + A$$
$$= 122.8 + (\alpha + \beta)$$
$$= 122.8 + (17.9 - 6.9)$$
$$= 133.8 \text{ (observed 132.7)}.$$

(b) $CH_2{}^iC(CH_3)CONH_2$:

$$\delta^i = 122.8 + A = 122.8 + (\alpha + \alpha)$$
$$= 122.8 + (13.4 + 9.6)$$
$$= 145.8 \text{ (observed 139.2)}.$$

Table 4.12. Increments for A for the general structure $XCH(\alpha)=CH(\beta)$

Substituent X	Increment α	Increment β	Substituent	Increment α	Increment β
—CH₃	13.4	−6.9	—N≡⁺	18.9	−10.6
—Et	17.9	−9.5	—CH₂Y	12	−5
—Pr	15.7	−8.4	—CH₂COOH	6.9	−4.6
—ᵗBu	26.9	−13.0	—CH₂CN	5.9	−2.1
—CN	−15.6	15.1	—COOR	6.0	8.0
—Ph	13.0	−10.5	—COCH₃	14.9	6.7
—OCH₃	31.0	−38.2	—CHO	15.8	14.8
—OCOCH₃	18.9	−26.4	—CON<	9.6	3.1
—NHCOR	7.2	−28.5	—Cl	3.3	−5.6
—NO₂	22.8	−0.4			

*Y = an electronegative atom or group.

(c) (E)—$CH_3{}^iCH=CHCOOEt$:

$$\delta^i = 122.8 + A$$
$$= 122.8 + (13.4 + 8.0)$$
$$= 144.2 \text{ (observed 144.0)}.$$

Note that ^iC in the part structure (Z)-$RCH=CH^iC$— is shielded by between 4 and 9 ppm.

4.11.2.4 Alkynes

The δ value of a carbon atom ^iC can be calculated as the sum of a series of constants such that

$$\delta^i = 72.0 + A$$

where 72.0 is the δ value for ethyne. Increments for A are given in Table 4.13.

Table 4.13. Increments for A for the general structure $XC(\alpha)\equiv C(\beta)$

Substituent	Increment α	Increment β
—CH₃	7.0	−6.0
—Et	12.0	−3.5
—ⁱPr	16.0	−3.5
—Ph	12.5	6.5
—CH=CH₂	10.0	11.0
—CH₂OH	11.0	2.0
—COCH₃	31.5	4.0
—Cl	−12.0	−15.0

4.11.2.4.1 Examples

(a) $Bu^iC\equiv CCOCH_3$:

$$\delta^i = 72.0 + (\alpha + \beta)$$
$$= 72.0 + (12.0^* + 4.0)$$
$$= 88.0 \text{ (observed } 87.0).$$

*Et value taken as closest analogy.

(b) $p\text{-}ClC_6H_4{}^iC\equiv CCH_3$:

$$\delta^i = 72.0 + (12.5 - 6.0)$$
$$= 78.5 \text{ (observed } 79.6).$$

4.11.2.5 Benzenoid aromatics

The δ value of a carbon atom ^iC can be calculated as the sum of a series of constants such that

$$\delta = 128.5 + A$$

where A is the sum of the increments allowed for the substituents at positions 1, 2, 3 and 4 and 128.5 is the δ value for benzene. Increments for A are given in Table 4.14.

4.11.2.5.1 Examples.

(a) ROC(O)— [benzene ring with NO$_2$ groups at positions 3 and 5, i at position 1]

$$\delta = 128.5 + A_1 + 2A_3$$
$$= 128.5 + 2.1 + (2 \times 0.9)$$
$$= 132.4 \text{ (observed } 132.2).$$

(b) NH_2—[benzene ring]—i

$$\delta = 128.5 + A_4$$
$$= 128.5 - 9.5$$
$$= 119.0 \text{ (observed } 119.0).$$

Table 4.14. Increments for A for the general structure

Substituent X	Increment			
	A_1	A_2	A_3	A_4
—CH_3	9.3	0.8	0.0	−2.9
—Et	15.8	−0.4	−0.1	−2.6
—^{i}Pr	20.3	−1.9	0.1	−2.4
—^{t}Bu	22.4	−3.1	−0.2	−2.9
—CH_2CO_2R	6.0	0.1	0.9	−1.4
—$CH=CH_2$	7.6	−1.8	−1.8	−3.5
—$C\equiv CH$	−6.1	3.8	0.4	−0.2
—C_6H_5	13.0	−1.1	0.5	−1.0
—CHO	8.6	1.3	0.6	5.5
—$COCH_3$	9.1	0.1	0.0	4.2
—CO_2H	2.1	1.5	0.0	5.1
—CO_2^-	7.6	0.8	0.0	2.8
—CO_2R	2.1	1.2	0.0	4.4
—$CONH_2$	5.4	−0.3	−0.9	5.0
—CN	−15.4	3.6	0.6	3.9
—Cl	6.2	0.4	1.3	−1.9
—OH	26.9	−12.7	1.4	−7.3
—OCH_3	31.4	−14.4	1.0	−7.7
—OC_6H_5	29.1	−9.5	0.3	−5.3
—$OCOCH_3$	23.0	−6.4	1.3	−2.3
—NH_2	18.7	−12.4	1.3	−9.5
—$N(CH_3)_2$	22.4	−15.7	0.8	−11.8
—NO_2	20.2	−4.8	0.9	5.8
—NHCOR	10.0	7.7	0.4	−4.1
—SH	2.2	0.7	0.4	−3.1
—SO_3R	7.0	1.0	−0.5	5.6
—SO_3H	15.0	−2.2	1.3	3.8
—CH_2X*	11.0	−0.5	0.0	−0.5

*X = an electronegative atom or group. The data shown in Tables 4.10, 4.11, 4.12 and 4.14 were adapted with permission from *Spectral Data for Structure Determination of Organic Compounds* by E. Pretch, T. Clerc, J. Seibl and W. Simon. Springer-Verlag (1983). Berlin, Heidelberg, New York and Tokyo.

4.11.2.6 Aldehydes and ketones (saturated and α, β-unsaturated)

The δ value of the carbonyl carbon atom can be calculated as the sum of a series of constants such that

$$\delta^i = 193.0 + A$$

where A is the sum of the increments for substituents in the α, β and γ positions and 193.0 is an assumed δ value for methanal. Increments for A are given in Table 4.15.

Table 4.15. Increments for A for the general structure, $-C_\gamma C_\beta C_\alpha{}^i\overset{O}{\overset{\|}{C}}C_\alpha C_\beta C_\gamma-$

Substituent	Increment		
	A_α	A_β	$A_\gamma{}^*$
—C(sp^3)	6.5	2.6	1.0
—C$_6$H$_5$	−1.2	0.0	0.0
—CH=CH$_2$	−0.8	0.0	0.0
—2-furanyl	−12.0	0.0	0.0

*When $C_\alpha C_\beta =$ CH=CH—.

4.11.2.6.1 Examples

(a) CH$_3$CH=CH^{i}CHO: $\delta^i = 193.0 + (\alpha + \gamma)$
$ = 193.0 - 0.8 + 1.0$
$ = 193.2$ (observed 192.4).

(b) CH$_3{}^i$CCH(CH$_3$)$_2$: $\delta^i = 193.0 + (2\alpha + 2\beta)$
$\|$
O
$ = 193.0 + 13.0 + 5.2$
$ = 211.2$ (observed 212.1).

4.11.2.7 Carboxylic acids and esters

The δ value of the carbonyl carbon can be calculated in the usual way from the equation

$$\delta^i = 166 + A$$

where 166 is the assumed δ value for methanoic acid. Increments for A are given in Table 4.16.

Table 4.16. Increments for A for the general structure $-C_\gamma C_\beta C_\alpha \overset{O}{\overset{\|}{C}}OX$

Substituent	Increment			
	α	β	γ	X
—C(sp^3)	11	3	−1	−5
—C$_6$H$_5$	6	1	0	−8
—CH=CH$_2$	5	0	0	−9

4.11.2.7.1 Examples

(a) $CH_3COCH_2CH_2{}^iCOCH_2CH_2CH_3$:
$$\underset{\underset{O}{\|}}{}$$

$$\delta^i = 166 + \alpha + \beta + X$$
$$= 166 + 11 + 3 - 5$$
$$= 175 \text{ (observed 172.7)}.$$

(b) $CH_2\!=\!CH^iCOOMe$:

$$\delta^i = 166 + \alpha + X$$
$$= 166 + 5 - 5$$
$$= 166 \text{ (observed 165.5)}.$$

4.11.2.8 Amides

The δ value for the amide carbon atom can be calculated in the usual way from the equation

$$\delta^i = 165 + A$$

where 165 is an assumed δ value for the methanamide. Increments for A are given in Table 4.17.

Note that the *syn* carbon atoms (designated α'') are shielded by 3–5 ppm with respect to the *anti* carbon atoms (α').

4.11.2.8.1 Example

$C_6H_5NH^iCOCH_2CH_2CH_3$:

$$\delta^i = 165 + \alpha + \beta + \alpha'$$
$$= 165 + 7.7 + 4.5 - 4.5$$
$$= 172.7 \text{ (observed 172.1)}.$$

Table 4.17. Increments for A for the general structure

$$-C_\beta C_\alpha C\underset{\underset{\|}{O}}{N}\!\!\diagdown_{C_{\alpha''}C_{\beta''}-}^{C_\alpha C_{\beta'}-}$$

Substituent	Increment				
	α	β	γ	α'	β'
—C(sp^3)	7.7	4.5	−0.7	−1.5	−0.3
—C$_6$H$_5$	4.7	—	—	−4.5	—
—CH=CH$_2$	3.3	—	—	—	—

Table 4.18. Effect of substituent in nitriles on δ values

Substituent	δ value	Substituent	δ value	Substituent	δ value
—Me	117.7	—CH_2Cl	115.7	—p-$C_6H_4NO_2$	112.2
—Et	120.8	—CH=CH_2	117.2	—2-furanyl	111.7
—^iPr	123.7	—C_6H_5	118.7	—cinnamyl	118.3

4.11.2.9 Nitriles

The δ^i value for the RC≡N nitrile-carbon atom is fairly constant. The examples in Table 4.18 indicate the way in which the δ value varies in different environments.

4.11.2.10 More complicated compounds

The use of the tables has so far been illustrated by reference to very simple compounds. Fair predictions (± 4 ppm) of δ values can be made for carbons in more complicated structures e.g. **4.13** and **4.14**.

(4.13)

(4.14)

4.11.2.10.1

(a) **4.13**:

$$\delta = -2.3 + A + B$$
$$= -2.3 + (\alpha^1 + \alpha^2 + \alpha^3 + 2\beta^1 + \beta^2 + 2\gamma^1) + (3° \rightarrow 3° + 3° \rightarrow 2° + 3 \rightarrow 2°)$$
$$= -2.3 + [9.1 + 22.0 + 28.3 + (9.4 \times 2) + 2.0 + (-2.5 \times 2)]$$
$$+ (-9.5 - 3.7 - 3.7)$$
$$= 56.0 \text{ (observed 59.8).}$$

(b) **4.14**:

$$\delta = -2.3 + (2\alpha^1 + 2\alpha^2 + 2\beta^1 + \beta^2 + \beta^3 + 5\gamma^1 + \gamma^2)$$
$$+ (4° \rightarrow 3° + 4° \rightarrow 2° + 4° \rightarrow 2° + 4° \rightarrow 1°)$$
$$= -2.3 (18.2 + 98.0 + 18.4 + 10.1 + 11.3 - 12.5 - 6.2)$$
$$+ (-15.0 - 8.4 - 8.4 - 1.5)$$
$$= 101.7 \text{ (observed 97.6).}$$

NUCLEAR MAGNETIC RESONANCE SPECTROSCOPY

The two examples shown above were taken at random. The unexpectedly close agreement between the calculated and observed values probably results from chance cancellation of errors.

4.11.2.11 Another application of correlation tables

In general, the more crowded or strained is the molecule and the more closely spaced are the functional groups, the less accurate are the values calculated using these tables.

Another approach to the calculation of ^{13}C δ values for the less predictable molecular types depends on the availability of values of some closely related structure. Here the tables can be used to make adjustments to the δ values observed for the related structure, for which we find poor correlation between calculated and observed values.

Let us consider an example in which we wish to predict the δ value for ^iC in the ketol **4.16**. The C–5 of the cyclopentenone **4.15** is observed at 34.1 ppm but is calculated to fall at 43.1 ppm, i.e. $-2.3 + (22.5 + 9.1) + (6.9 + 6.9) = 43.1$.

(**4.15**)   (**4.16**)

Using the observed value of 34.1, adjustments are made to the calculation of the δ value for C-5 in the ketol 4.16:

$$\delta^i C = 34.1 \text{ (reference value)} + 9.1 \text{ (new } \alpha) + 10.1 \text{ (new } \beta) + 9.4 \text{ (new } \beta)$$
$$+ 2 \times -2.5 \text{ (new } \gamma) + 2 \times -3.7 \text{ (new } 3° \rightarrow 2°)$$
$$= 50.1 \text{ (observed 47.4 ppm)}.$$

The calculation for, say, the carbinol carbon can be made in the usual way since this is not part of the five-membered ring system, i.e.

$$\delta^i C\text{—OH} = -2.3 + (49.0 + 9.1 + 9.1) + (9.4 + 9.4 + 9.4 + 3.0)$$
$$+ (2x - 2.1) + (2 \times -9.5)$$
$$= 72.0 \text{ (observed 72.9)}.$$

4.12 SIMPLIFICATION OF SPECTRA

4.12.1 High-field NMR

A spectrum taken at high field has a greater dispersion or spread of signals than a spectrum taken at low field. This is evident if we consider half of an AX system with a J value of 20 Hz. At 60 MHz the doublet spans one third (20/60) of 1 ppm.

whereas at 400 MHz it spans only one twentieth (20/400) of 1 ppm. Consequently, overlapping multiplets seen at low field are often well resolved at high field.

4.12.2 Spin decoupling

Another way of simplifying spectra is by use of double resonance techniques, one of which is spin decoupling. If the spectrum of a compound containing an AX system is taken by scanning with a weak radio-frequency in the usual way whilst nucleus X is continuously irradiated with a strong, second radio-frequency, hence double irradiation, the spin decoupling condition is met. The effect of the strong radio-frequency irradiation is to produce a rapid exchange between the two possible spin states of the X nucleus, so fast that the adjacent nucleus A senses not two but a single, time-averaged magnetic field from X. A then appears as a singlet, that is, as though nucleus X were no longer in the molecule. This is the effect of all spin-decoupling measurements, i.e. all the couplings from other nuclei to the strongly irradiated nucleus or nuclei are removed and a simplified spectrum is produced. By comparing the 'normal' and decoupled spectra, the nuclei which are decoupled and so in close through-bond proximity to the irradiated one can be

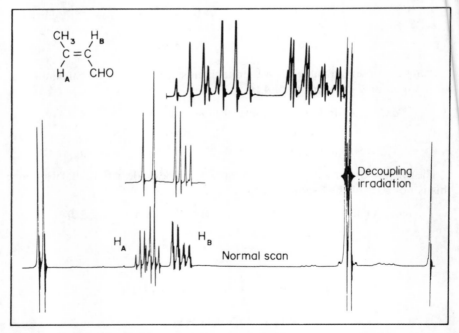

Figure 4.33 Reproduced from *NMR Quarterly*, 1973, No. 7, by permission of Perkin–Elmer Ltd. © Perkin–Elmer, 1973.

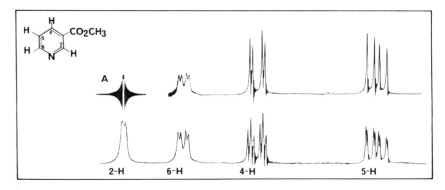

Figure 4.34 Reproduced from *NMR Quarterly*, 1972, No. 5, by permission of Perkin–Elmer Ltd. © Perkin–Elmer, 1972.

identified. An example of such a measurement is illustrated in Fig. 4.33 using crotonaldehyde as example.

The top trace shows the olefinic protons with a three-times scale expansion compared with the normal spectrum in the bottom trace. In the expanded trace six quartets can be seen with some overlapping of the left-hand pair. We interpret the 'quarteting effect' as being due to coupling to the methyl protons by both olefinic protons, and this is confirmed when that region of the spectrum is re-run with continuous irradiation of the methyl protons. This is shown in the middle, spin-decoupled trace which can now be described by 6.1 (1H, dd, $J = 6.5$ and 15.5 Hz), 6.9 (1 H, d, $J = 15.5$ Hz) and, but not shown, 9.5 (1 H, d, $J = 6.5$ Hz). Hence the effect of the methyl group is removed and the residual splitting is seen to be of the eight-line AMX type where A, the aldehydic proton is not coupled to the proton at 6.9 ppm. In more complicated examples the simplification brought about by a spin-decoupling experiment or a series of them often allows δ and J values to be determined which would otherwise not be accessible. A second example using methyl nicotinate is shown in Fig. 4.34. The lower trace shows the normal spectrum with the lowest field doublet assigned to the C-2 proton since it is deshielded by both nitrogen and the adjacent carbomethoxy group. Irradiation of this signal removes coupling to the C-4 and C-5 protons, as seen in the upper spectrum, leaving a typical AMX pattern, similar to example (c) in Table 4.6.

4.12.3 Decoupling by quadrupole effects

The lack of fine structure on NMR-active nuclei close to ^{14}N is discussed in Appendix 4.16.2. The same observation is made for such nuclei close to the halogens except fluorine. Chlorine, bromine and iodine each have more than one isotope possessing a non-zero, magnetic spin-quantum number but do not couple with ^{13}C atoms to which they are bound or to protons on adjacent carbons. As before, the explanation is that the quadrupole moments of the nuclei concerned cause rapid relaxation of the spin states and so decouple themselves from others nearby.

4.12.4 Lanthanide shift reagents

Sometimes overlapping multiplets can be resolved by the addition of shift reagents to the test solution. These are usually lanthanide organometallics (**4.17**) which have good solubility in NMR solvents. Although paramagnetic, they cause only moderate enhancements of relaxation rates and so limited line-broadening effects. The reagents, three of which are listed below, act as Lewis acids, interacting with basic centres in the molecule under observation. The strength of the interaction varies directly with the basicity of the centre (i.e. decreases in the order amines, alcohols, ketones, ethers, esters, nitriles) and with the concentration of the reagent. It varies indirectly with the polarity of the solvent and the temperature at which the spectrum is measured. Ideally, these measurements would be made in CCl_4 at low temperatures on a compound having only one strongly basic functional group. The protons closest to the basic centre are those most strongly affected, being shielded by praseodymium complexes and deshielded by europium complexes.

(**4.17**)

$Pr(dpm)_3: R_1 = R_2 = CMe_3$
$Eu(dpm)_3: R_1 = R_2 = CMe_3$
$Eu(fod)_3: R_1 = CMe_3, R_2 = n-C_3F_7$

A dramatic example of the effect of Eu $(dpm)_3$ on the spectrum of heptan-1-ol is illustrated in Fig. 4.35. The methylene protons are almost completely resolved under the conditions described.

Although the use of high-field spectroscopy renders unnecessary many shift reagent measurements, they are still useful where overlapping multiplets remain unresolved at high field and are indispensable for the assessment of enantiomeric ratios (see Section 4.13.3).

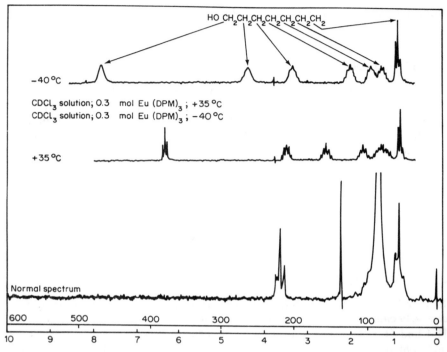

Figure 4.35 Effect of a lanthanide shift reagent on the spectrum of heptan-1-ol. Reproduced from *NMR Quarterly*, 1971, No. 1, by permission of Perkin–Elmer Ltd. © Perkin–Elmer 1971.

4.12.5 Deuteration

See Section 4.14.2.

4.12.6 Nuclear Overhauser effect

The application of the nuclear Overhauser effect (nOe) does not strictly simplify spectra, but does simplify the interpretation of spectra. It is a side-effect of double resonance techniques, as briefly outlined in Section 4.12.2, which causes an increase in the intensities observed for resonance signals due to nuclei close to the doubly irradiated nucleus. The intensity-enhanced signals need not be due to nuclei closely spaced through bonds. They may in fact be separated by many tens of bonds, but be coupled through space by being brought close together, generally within 4 Å, by some geometrical constraint or conformational preference. It is the through-space enhancement effect which may provide valuable information on the spatial relationship of atoms which are distant through bonds.

4.12.6.1 Example 1

Both geometrical isomers of methyl-3-iodocrotonate (**4.18**) were isolated from a reaction mixture. It was not clear from their proton spectra which was which. However, when for each compound the olefinic proton was continuously irradiated, only one isomer showed an enhancement in intensity of the higher field methyl group. This must be the Z isomer.

No enhancement ⟶ Me, CO_2Me 20% enhancement ⟶ Me, H ⟵ irrad
                (E) H ⟵ irrad   (Z) CO_2Me

(**4.18**)

4.12.6.2 Example 2

In an nOe measurement, irradiation at the H-1 signal position of the vinylpiperidine **4.19** caused enhancement of the signal intensity for H-2 and H-3 but not for H-4 and H-5. The three chiral carbon centres possessed by the molecule are highlighted. In a separate nOe measurement, irradiation at H-4 was found to cause enhancement of H-5. These results confirmed the full relative stereochemistry depicted in **4.19**.

(**4.19**)

4.13 EFFECT OF CHIRALITY ON THE NMR SPECTRUM

4.13.1 Non-equivalent protons

The constituent enantiomers of a racemic mixture have identical properties in an achiral environment, so they are indistinguishable under the normal conditions of the NMR experiment. Where a methylene group is located next to or near a chiral centre, e.g. RCH_2—CXYZ, rotations about the carbon—carbon bond give, as usual, three conformational energy minima. None of the three corresponding conformations has a plane of symmetry, so the individual protons of the methylene group are, despite rapid rotations, always in different chemical and magnetic environments. Therefore, provided that $\Delta v/J$ is large, for which condition at low field an anisotropic group must be nearby, an AX spectrum results.

The three staggered conformations of RCH_2—$CXYZ$

In example (h) in Table 4.4, the chiral centre is β to the methylene protons and the amide bond acts as the anisotropic group. If one of X, Y or Z is a proton, then we have, depending on the $\Delta v/J$ ratios of the three interacting nuclei, an AMX or ABX system as discussed in Section 4.7.2.3.

4.13.2 Non-equivalence in achiral compounds

Compounds of the type RCH_2CHXCH_2R, even though not chiral, do possess diastereotopic methylene groups. That is, if either one or the other proton of a given methylene group is replaced by a new substituent, the two products so formed will be diastereomers. This is the easiest test for non-equivalence of such protons. A second way is that applied above to compounds having the —CH_2CXYZ part structure where, for **4.22** and **4.23**, rotations about the C—O or C—N bonds never produce conformers having a plane of symmetry. Chlorocyclohexane (**4.20**) and cyanocyclopropane (**4.21**), in addition to the acetals and amine salts already referred to contain non-equivalent, diastereotopic methylene groups.

(4.20)   (4.21)

(4.22)   (4.23)

The spectrum of **4.22** is reproduced in Fig. 4.36, in which the eight-line multiplet between 3.9 and 4.7 ppm results from the basic AB quartet being further split into eight by coupling to the proton on nitrogen.

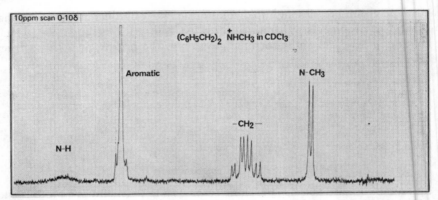

Figure 4.36 Spectrum of compound **4.22**. Reproduced from *NMR Quarterly*, 1973, No. 7, by permission of Perkin–Elmer Ltd. © Perkin–Elmer, 1973.

4.13.3 Chiral shift reagents

Polarimetry has traditionally been used to estimate the efficiency of enantiomeric resolutions, but the method depends on a knowledge of the rotation of one of the enantiomers. The use of chiral shift reagents (see Section 4.12.4) such as tris (d, d-dicamphoylmethanato) europium (**III**) (**4.24**), Eu(dcm)$_3$, often provides a more convenient analysis. If a chiral shift reagent, say of R configuration, is dissolved in a mixture of enantiomers of a compound containing a basic functional group, then an R–R and R–S interaction will take place between them. Because this produces a pair of diastereomeric complexes, each will give rise to a different spectrum and any enantiomeric excess can be calculated by comparison of the integral heights run over like protons in each diastereomer.

(**4.24**)

The spectrum of a 40:60, R–S mixture of enantiomers of 1-phenyl-ethylamine after addition of Eu(dcm)$_3$ is shown in Fig. 4.37b and the spectrum of the pure R enantiomer containing a similar amount of shift reagent is shown in Fig. 4.37a. Note that the reagent–amine interaction causes downfield shifts directly related to the distance of the shielded protons from the site of complexation, giving the order methine, methyl, *ortho*, *meta* then *para* protons.

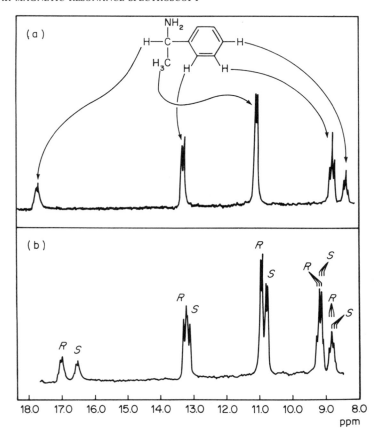

Figure 4.37 Spectra of solutions prepared from (a) (S)-α-phenylethylamine (10 μl) and (b) a mixture of (R)- and (S)-α-phenylethylamine (7 and 5 μl, respectively), in carbon tetrachloride solution. The chemical shift scale applies only to the spectrum of the mixture (b); That of the pure S enantiomer (a) was displaced slightly to lower field owing to differences in concentrations of the samples. Reprinted with permission from *J. Am. Chem. Soc.*, **92**, 6980 (1970), G. M. Whitesides and D. W. Lewis. American Chemical Society.

4.14 DYNAMIC EFFECTS OBSERVED BY NMR

4.14.1 Proton exchange

Spectrum (d) in Fig. 4.38 is the 'normal' spectrum obtained for methanol in which there is no coupling observed between the methyl and the hydroxy protons. On

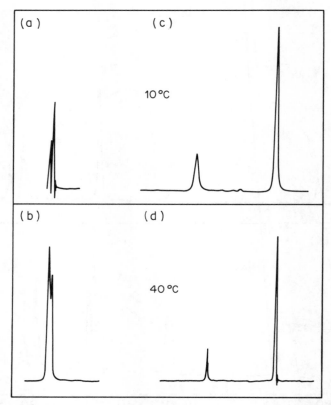

Figure 4.38 Spectra of methanol at different temperatures. Reproduced from *NMR Spectroscopy*, 1983 by Harald Günther with permission of Georg Thieme Verlag, Stuttgart.

reflection this is surprising since this coupling is ^3J, i.e. only through three bonds, and so the effect of a strong coupling should normally be seen as in the AX_3 spectrum (a). The difference in the appearance of these spectra is due to the difference in the rate of exchange of the hydroxylic proton at the different temperatures. At $-65°C$ the rate of exchange is slow on the NMR time scale so that the proton is resident on the oxygen atom long enough for the methyl group to sense its two possible spin states and for the proton to sense the four spin state combinations of the methyl group. At 40°C the exchange is fast on the NMR time scale so that the two kinds of protons appear to be independent of each other. Intermediate situations, where the lifetime of the O—H bond is of the same order as the time of the NMR experiment (*ca* 0.16 s) are depicted in (b) and (c), where different degrees of line broadening are evident.

A similar series of spectra (a)–(d) can be obtained by successive additions of acid to a very pure sample of methanol. In this case only a slow exchange of protons occurs in the pure sample, the rate of exchange increasing with increasing amounts of acid which catalyse the reaction. The above phenomena are observed for most alcohols. Incidentally, the spectra of sugars (O—H protons) and peptides (amide N—H protons) run in DMSO-d_6 always show the appropriate multiplicities for their ^3J couplings. This is due to the strong solvation of these polar bonds by DMSO, which hinders proton exchange. Thus each OH group can be identified as primary, secondary or tertiary and each NH as being part of a glycine residue (triplet) or other amino acid residue (doublet).

4.14.2 Deuterium exchange

Protons attached to oxygen, nitrogen and sulphur readily undergo exchange with deuterium. This property is used in NMR spectroscopy to confirm their presence in the functional groups shown in Appendix 4.16.2. This is achieved by running a normal spectrum then adding one drop of D_2O to the sample, shaking it and re-running the spectrum. The acidic protons are replaced by deuterium atoms and so disappear from the spectrum, leaving a signal due to HOD at 4.8 ppm.

Particularly acidic protons attached to carbons, as for example in β-keto esters, 1, 3-diketones, etc., are also replaced in this way, usually more slowly than in the above examples, by participation of D_2O in the keto–enol interconversion process.

As explained in Section 4.14.1, sharp absorptions and H—C—X—H couplings are observed for acidic protons in DMSO-d_6 solution, but in most other solvents their sharpness and δ values depend on the extent of hydrogen bonding in which they are involved and so are related to the concentration of the sample and the degree of steric crowding about the acidic proton. If more than one type of acidic proton is present, rapid exchange between them often gives rise to one broad signal at an averaged position.

4.14.3 Ring inversion processes

The ring inversion between the stable chair forms of cyclohexane via successive skew boat, boat and skew boat conformations is the most obvious example of this phenomenon. At room temperature cyclohexane gives rise to a fairly broad singlet absorption at 1.3 ppm, indicating that inversion takes place rapidly compared with the NMR time scale. This is the time-averaged spectrum. At $-100°C$ the rate of inversion is slowed so that protons remain in the axial or equatorial positions long enough for the spectroscopic method to detect them at their individual δ values when $\delta_{eq} - \delta_{ax} = 0.5$ ppm.

The δ value observed at room temperature is mid-way between those seen at $-100\,°C$. This is because the time-averaged δ (and J) values are 'weighted means' of the individual conformers, which we may regard as a 50:50 mixture of the two

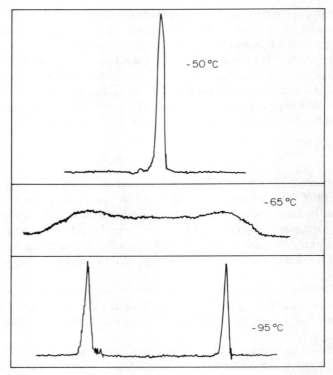

Figure 4.39 Spectra of undecadeuteriocyclohexane at −50, −65 and −95 °C.

chair forms. An interesting variation on the above is the case of undecadeuteriocyclohexane (**4.25**). In this instance the chair conformations are different and the singlet spectrum due to each of these is seen in Fig. 4.39. The conformer with the axial proton resonates at 1.12 ppm, shielded with respect to the equatorial proton (1.60 ppm). Both compounds give rise to broadened singlets (6 Hz), the broadening being due to small couplings to deuterium. These may be sharpened by deuterium decoupling.

(4.25)

At intermediate temperatures line broadening is observed. The temperature at which the lines due to individual conformers merge into broad absorptions is termed the coalescence point, and from this temperature the energy barrier to inversion can be calculated. In the case of cyclohexane and many of its simple derivatives this value is near 45 kJ mol^{-1}. An expected exception is the cis-1, 2-di-tert-butyl derivative, which has a barrier to inversion of 68 kJ mol^{-1} corresponding to a coalescence temperature of 35 °C; it is expected since the energy barrier to conversion in this derivative is enhanced by a *tert*-butyl–*tert*-butyl eclipsing interaction.

4.14.3.1 Inversion of aziridines

Inversion of aziridines such as the *N*-methyl derivative (**4.26**) involves the surmounting of an energy barrier of about 80 kJ mol^{-1} at 115 °C. This is larger than that normally encountered for amine inversions owing to the resistance of the three-membered ring to the ideal 120° bond angle encountered in the normal inversion transition state. Below 115 °C there are two sets of ring protons and above that temperature one, time-averaged, set.

(**4.26**)

4.14.4 Rotation about bonds

4.14.4.1 *N,N*-Dimethyl-*tert*-butylamine

The spectrum of *N,N*-dimethyl-*tert*-butylamine (**4.27**), shown in Fig. 4.40a, appears at ordinary temperatures as two sharp singlets which broaden around −150 °C. As the temperature is lowered further so the higher field singlet broadens and in spectrum (b), taken at ca −165 °C, becomes resolved into two singlets with an intensity ratio of 2:1.

(**4.27**)

The molecule has three identical staggered conformations with a symmetry plane such that only two of the three methyl groups are identical. At ordinary temperatures fast rotation about the C—N bond gives a time-averaged spectrum

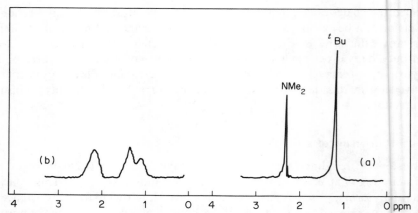

Figure 4.40 Spectrum of N, N-dimethyl-*tert*-butylamine at (a) 100 °C and (b) −165 °C.

in which the three *tert*-butyl methyl groups experience identical averaged environments. At −165 °C the rate of rotation is very slow compared with the NMR time scale and in effect a frozen staggered conformation (**4.27**) is observed, giving the 2:1 ratio of high-field methyl groups.

Where more crowded C—N and C—C bonds are concerned the coalescence temperature may be at or above room temperature, so that simplification of the spectrum by increasing the rate of rotation about the bond in question may be achieved by heating the sample.

4.14.4.2 Amide bonds

[(**Z**)-**4.28**]

[(**E**)-**4.28**]

The spectrum of N-benzyl-N-methylformamide (**4.28**) might be expected to consist of four singlets in the ratio 1:5:2:3, and indeed this is the case if the spectrum is taken at elevated temperatures. At room temperature, however, the

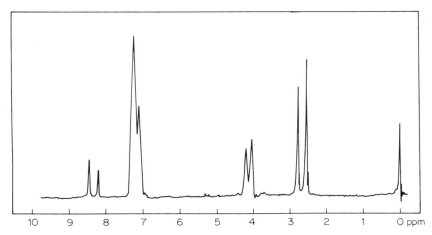

Figure 4.41 Spectrum of N-benzyl-N-methylformamide.

spectra of both E and Z conformers are observed (Fig. 4.41). The double bond character of the C—N bond implied by the charge-separated canonical forms serves to raise the activation barrier to rotation to ca 92 kJ mol^{-1}, so that fast rotation and time averaging of signals does not take place until a temperature of 160 °C, the coalescence temperature being taken as 127 °C.

A similar observation is made with dimethylformamide but a difference is that the rotation barrier, in this case 85 kJ mol^{-1}, separates identical conformers, so the signal intensities for the methyl signals are equal. In the former case the corresponding, resolved signal intensities maintain a constant E:Z ratio (5:4), which gives a direct measure of population distribution at that temperature and so the difference in energy between the E and Z isomers.

Similar rotational effects are observed about C—O bonds in esters and indeed any conjugated system such as $C_6H_5X=Y$ and $C_6H_5\ddot{X}$— where resonance effects set up barriers to rotation.

4.15 PROBLEMS

1. Analyse the ^{1}H NMR spectra (a), (b) and (c) in Fig. 4.42 in terms of the structures provided.

2. Predict the multiplicity of the signals due to the underlined nuclei in the following structures: (a) C$\underline{H}_3$Cl; (b) C$\underline{H}_3$F; (c) PhC$\underline{H}_2$Br; (d) (C$\underline{H}_3$CH$_2$)$_2$O; (e) PhC$\underline{H}_2$COOR; (f) PhC$\underline{H}_2$CHO; (g) PhC$\underline{H}_2$CH$_2$Ph; (h) PhC$\underline{H}_2$PPh$_2$; (i) PhC$\underline{H}_2$CHFPh; (j) PhC$\underline{H}_2$SH; (k) PhC$\underline{H}_2$OH (in DMSO); (l) ^{13}C$\underline{H}$Cl$_3$; (m) PhC$\underline{H}_2$SOCH$_3$; (n) PhC$\underline{H}$F$_2$; (o) ^{13}C$\underline{D}_3$SOCD$_3$; (p) ^{13}C$\underline{D}$Cl$_3$.

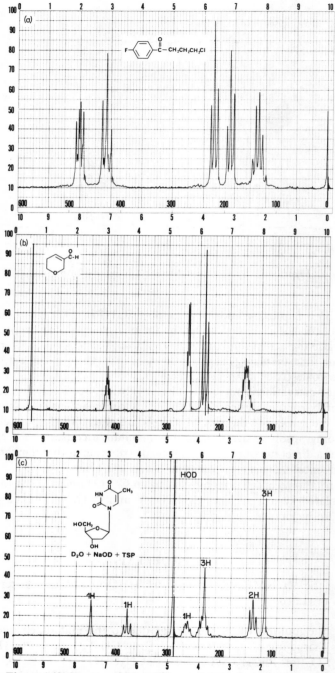

Figure 4.42 Spectra of (a) γ-chloro-*p*-fluorobutyrophenone, (b) 5,6-dihydro-2*H*-pyran-3-carboxaldehyde and (c) thymidine. Reprinted with permission of Aldrich Chemical Co., Inc.

Explain the statements in 3–7 below:

3. The proton spectrum of 1,3,5-trioxacyclohexane run at 35 °C consists of a singlet and at −85 °C an AB quartet, $J = 12$ Hz; the carbon spectrum contains only a triplet, $J = 162$ Hz, at the former temperature but a doublet of doublets at the latter.

4. There are five lines in the fully decoupled carbon spectrum of benzaldehyde determined at room temperature and seven lines at −135 °C.

5. The ^{1}H NMR spectrum of methanamide consists of two broad lines when run at 100 °C and three such lines below 0 °C. If the low-temperature spectrum is taken with decoupling from nitrogen, three sets of four lines (doublets of doublets) appear.

6. The aromatic region of the ^{1}H NMR spectrum of N-methyl-2,4,6-trinitroaniline comprises an AB system, $J = 2.0$ Hz, when run at −80 °C and an A_2 singlet at +50 °C.

7. Four lines only, in an approximate ratio 1:10:5:10, appear in the aromatic region of the fully decoupled carbon spectrum of N-methylaniline measured at room temperature. If the temperature at which the measurement is made is lowered to near −100 °C, six lines are observed in an approximate ratio 1:5:5:5:5:5.

4.15.1 Answers to problems

1. (a) δ 2.2, —CH$_2$—; 3.15, —CH$_2$CO—; 3.7, —CH$_2$Cl; 7.1, Arom 2H *ortho*-F (note $^3J_{F-H} = {^3J_{H-H}} = 8$ Hz); 8.0, Arom 2H *ortho*-CO, $^4J_{F-H} = 6$ Hz.
   (b) δ 2.5, —CH$_2$—C=(^3J, ^3J and ^5J coupled); 3.8, —CH$_2$O; 4.3, O—CH$_2$—C=; 6.95, =CH—(^3J and ^4J coupled); 9.4, —CHO.
   (c) δ 1.9, —CH$_3$; 2.25, 2-CH$_2$ sugar; 3.8, 4-CH and 5-CH$_2$ sugar; 4.4, 3-CH sugar; 6.35, 1-CH sugar; 7.5, =CH—.

2. (a) s; (b) d, q; (c) s; (d) t, q; (e) s; (f) d, t; (g) s; (h) d, t; (i) dd, dt, dt; (j) d, t; (k) d, t; (l) d, d; (m) AB q; (n) t, d; (o) sept; (p) t, d.

3. The chair forms are 'frozen' at −80 °C.

4. At low temperature there is no rotation about the Ar—C bond, hence no equivalence of carbon atoms.

5. Below 0 °C there is no rotation about the C—N bond. Broadening of signals is due to short lifetime of nitrogen spin states (quadrupole broadening). Complete decoupling removes effect of ^{14}N and leaves an AMX system.

6 and 7. Similar cases to 4.

4.16 APPENDICES

4.16.1 Chemical shift values

4.16.1.1 Aldehydes, 9.0–10.5 ppm

Aldehydic protons usually appear around 9.8 ppm. Benzaldehyde (10.0 ppm) derivatives having $+M$ substituents tend to higher field, e.g. p-MeOC$_6$H$_4$CHO (9.9 ppm), and those with $-M$ substituents to lower field, particularly when located *ortho* to the CHO group, e.g. o-NO$_2$C$_6$H$_4$CHO (10.5 ppm). Aliphatic aldehydes are shielded if crowded, e.g. (CH$_3$)$_3$CHO (9.5 ppm), with respect to butanal (9.7 ppm).

4.16.1.2 Alkynes, 1.3–3.4 ppm

Alkynic protons normally occur around 2.2 ppm. Conjugation with alkenes, e.g. HC≡CC(Me)=CH$_2$ (2.88 ppm), and aromatics, e.g. HC≡CC$_6$H$_5$ (3.05 ppm), particularly when $-M$ substituents extend the conjugation, causes deshielding from this value. A similar but smaller effect is seen when the γ-carbon is substituted with $-I$ groups, e.g. HC≡CCH$_2$OH (2.5 ppm).

4.16.1.3 Oxiranes, aziridanes and cyclopropanes

These appear at 2.54, 1.61 (NH, 0.9) and 0.22 ppm, respectively, shifted upfield by about 1 ppm with respect to their larger ring or acyclic analogues.

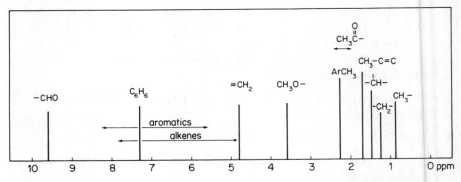

Figure 4.43 Typical chemical shift positions of commonly encountered proton types.

4.16.1.4 Alkanes

Estimation of δ values for substituted methanes where

$$\delta = 0.23 + A + B \text{ ppm}$$

A is the sum of the increments allowed for each α- substituent, and B equals the sum of the increments allowed for each β-substituent and 0.23 is the chemical shift of methane (Table 4.19).

Table 4.19. Increments for A and B for substituted methanes

Substituent	A	B	Substituent	A	B
—Alkyl	0.50	0.0	—NO$_2$	3.60	0.7
—C=C—	1.35	0.1	—COR	1.70	0.3
—C=C—	1.45	0.1	—COOR	1.60	0.2
—C≡N	1.70	0.4	—CONR$_2$	1.70	0.2
—Ph	1.85	0.2	—F	3.45	0.2
—OH	2.55	0.3	—Cl	2.55	0.3
—OR	2.35	0.3	—Br	2.35	0.4
—OPh	3.20	0.4	—I	1.80	0.5
—OCOR	3.15	0.4	—CF$_3$	1.15	0.2
—NR$_2$	1.60	0.1	—SR	1.65	0.3
—NHCOR	2.30	0.2	—SO$_2$R	2.20	0.4
—N$^+_\leqslant$	2.30	0.4			

4.16.1.4.1 Examples

1. $CH_3C^iH_2CH_2NO_2$:

$$\delta = 0.23 + 2A + B$$
$$= 0.23 + (2 \times 0.5) + 0.7$$
$$= 1.93 \text{ (observed 2.07)}.$$

2. $CH_3C^iH(SH)CH_3$:

$$\delta = 0.23 + 2A + B$$
$$= 0.23 + (2 \times 0.5) + 1.65$$
$$= 2.88 \text{ (observed 3.20)}.$$

3. ClC^iH_2COOR:

$$\delta = 0.23 + 1.6 + 2.55$$
$$= 4.38 \text{ (observed 4.05)}.$$

4.16.1.5 Alkenes

$$\begin{array}{c}
\text{trans } R_1 \diagdown \quad \diagup R_3 \text{ gem} \\
C=C \\
\text{cis } R_2 \diagup \quad \diagdown H
\end{array}$$

(**4.29**)

Table 4.20. Increments for A for substituted alkenes

Substituent	gem	cis	trans
—Alkyl	0.50	−0.25	−0.30
—C=C—	1.00	−0.05	−0.20
—C≡C—	0.50	0.35	0.10
—C≡N	0.25	0.80	0.60
—Ph	1.35	0.4	−0.10
—OR	1.15	−1.05	−1.30
—OCOR	2.10	−0.40	−0.65
—NR_2	0.70	−1.20	−1.30
—COR	1.10	1.15	0.80
—COOR	0.90	1.20	0.60
—$CONR_2$	1.40	0.90	0.35
—Cl	1.00	0.20	0.05
—Br	1.05	0.40	0.55
—I	1.30	0.70	0.60
—SR	1.00	−0.25	−0.05
—SO_2R	1.60	1.15	0.95

Estimation of δ values for substituted alkenes **4.29** where

$$\delta = 5.28 + A \text{ ppm}$$

A is the sum of the increments allowed for each of the substituents and 5.28 is the chemical shift of ethene (Table 4.20).

4.16.1.5.1 Examples

1. (E)—$CH_3(I)C=C^iHCOOR$:

$$\delta = 5.28 + 0.9 + (-0.3) + 0.70$$
$$= 6.58 \text{ (observed 6.57)}.$$

2. (Z)—$HC\equiv CC^iH=CHOCH_3$:

$$\delta = 5.28 + 0.50 + (-1.30)$$
$$= 4.48 \text{ (observed 4.52)}.$$

4.16.1.6 Aromatics

Estimation of δ values for substituted benzenes where

$$\delta = 7.27 + A \text{ ppm}$$

A is the sum of the increments allowed for each of the substituents and 7.27 is the chemical shift of benzene (Table 4.21).

Table 4.21. Increments for A for substituted aromatics

Substituent	ortho	meta	para
—Alkyl	−0.50	−0.05	−0.20
—C=C—	0.05	−0.05	0.00
—C≡C—	0.15	0.00	0.00
—C≡N	0.30	0.20	0.30
—Ph	0.35	0.15	0.10
—OH	−0.50	−0.05	−0.45
—OR	−0.45	−0.10	−0.45
—OPh	−0.30	−0.05	−0.25
—OCOR	−0.20	0.00	−0.10
—NH_2	−0.75	−0.25	−0.65
—NR_2	−0.65	−0.20	−0.65
—NHCOR	0.40	−0.10	−0.30
—NO_2	0.95	0.20	0.35
—COR	0.65	0.10	0.30
—COOR	0.75	0.10	0.20
—$CONR_2$	0.60	0.10	0.20
—F	−0.30	0.00	−0.20
—Other hal.	0.20	0.10	0.05
—SO_2R	0.70	0.30	0.40

The data shown in Tables 4.19, 4.20 and 4.21 were adapted with permission from *Spectral Data for Structure Determination of Organic Compounds* by E. Pretch, T. Clerc, J. Seibl and W. Simon. Springer-Verlag (1983), Berlin, Heidelberg, New York and Tokyo.

4.16.1.6.1 Examples

1. Br—C$_6$H$_3$(${}^iH$)—NO_2:

$\delta = 7.27 + 0.95 + 0.1$
$= 8.32$ (observed 8.38).

2. MeO—C$_6$H$_3$(${}^iH$)—CH=CHMe:

$\delta = 7.27 − 0.45 − 0.05$
$= 6.77$ (observed 6.80).

4.16.2 Acidic Protons

The range of δ values and the J values for acidic protons in different functional groups are given in Table 4.22.

Table 4.22. Range of δ values and J values for acidic protons in different types of compounds

Functional group	δ range (ppm)	J (Hz)
Alcohols	0.5–5.0	~4–7
Phenols	5.0–8.5	
Enols	12–16	Broad or very broad
Carboxylic acids	10–13	
Thiols, alkyl	1–2	~8
Thiols, aryl	2–4.5	
Amines, alkyl	0.5–4.0	
Amines, aromatic	2.5–5.0	
Amides	5–10	Often broad, alkyl to higher field than aryl, primary to higher field than secondary, than tertiary. ~2–7
Ammonium $^+$NH	7–8	

4.16.2.1 N–H Coupling

Although N has $I = 1$, coupling to an attached hydrogen atom is not often observed owing to the nitrogen's quadrupole moment. This usually causes a fast exchange between its three possible spin states, thus nitrogen effectively decouples itself from nearby nuclei. If the rate of exchange is fast, complete 'decoupling' results; if it is close to the time of the NMR experiment, then line broadening results as with many amides, but if it is slow on the NMR time scale, then N–H coupling is observed. Amine-salt nitrogens fall into this last category, particularly the more symmetrical types. The N–H proton appears near 8 ppm as a widely spaced (ca 50 Hz) triplet with further couplings to vicinal ($^3J \approx 0.5$ Hz) and four-bond separated ($^4J \approx 2$ Hz) protons.

4.16.3 Typical proton–proton coupling constants

A value x is used to denote the number of bonds through which the coupling is carried, as in ^xJ.

4.16.3.1 Alkanes

Typical values for the system —^{1}CH$_2$^{2}CH$_2$^{3}CH$_2$^{4}CH$_2$— are $^2J_{1,1(gem)} = 14$, $^3J_{1,2(vic)} = 8$, $^4J_{1,3} = 0$–1.5 and $^5J_{1,4} = 0$ Hz. The geminal, ^2J, value varies (8–20 Hz). Electronegative substituents reduce the magnitude of J but substitution by alkenes or aromatics increases it.

The vicinal, ^3J, value is typically around 8 Hz in freely rotating systems but varies (0–16 Hz) for constrained systems. This latter case is often met with in cyclic compounds when conformational and rotational restrictions prevent

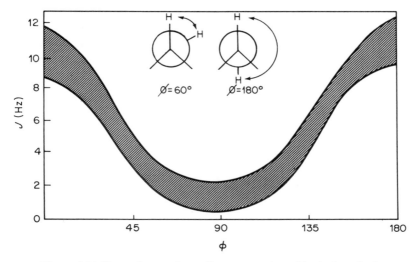

Figure 4.44 Dependence of coupling constant on dihedral angle ϕ.

averaging of δ and J values. The dependence of J on the dihedral angle ϕ is given approximately by the simple form of the Karplus equation, $J = 10 \cos \phi$.

The magnitude of J is also affected by torsional and ring strain and by the proximity of electronegative atoms, hence the latitude indicated in the graphical representation of the J–ϕ relationship (Fig. 4.44).

Longer range couplings, ^4J and ^5J, are more often observed in rigid molecules such as the bicyclic ketone **4.30** and more particularly where the W stereochemistry exists as indicated in the structure. Note here that in all three cases coupling occurs through four bonds and that in one an oxygen atom and in another a carbonyl carbon helps transmit the spin information. In each instance $J \approx 1.5$ Hz.

(4.30)

4.16.3.2 Alkenes

Typical values for the system

are $^2J_{1,1} = 2.5$, $^3J_{1,2} = 5$–14 (*cis*) and 12–18 (*trans*), $^4J_{1,3} = 0$–3 and $^3J_{2,3} = 5$ Hz.

For cycloalkenes the *cis* coupling constant decreases in roughly linear fashion, from 13 to 1 Hz, as the ring size decreases from seven to three.

The value of $^5J_{1,4}$ is 2 Hz for $-^1CH_2-^2CH=^3CH-^4CH_2-$. The homo-

(4.31)

allylic coupling is usually small but may be in the range 4–9 Hz in some fixed cyclic systems, e.g. compound **4.31** ($^5J_{1,4}$). For

$^3J_{1,2} = 10$ Hz.

4.16.3.3 Alkynes

For $H^1C\equiv{}^2C-^3CH$, $^4J_{1,3} = 2$ Hz and for $H^1C-^2C\equiv{}^3C-^4CH$, $^5J_{1,4} = 2$ Hz.

4.16.3.4 Aldehydes

Values of J are 2–3 Hz for HC—CHO and 8 Hz for C=CH—CHO.

4.16.3.5 Aromatics

Typical values for aromatic compounds are given in Table 4.23.

Table 4.23. Coupling constants for aromatic compounds

Compound	δ (ppm)	^3J (Hz)	^4J (Hz)	^5J (Hz)
Benzene		8 ($J_{1,2}$)	2 ($J_{1,3}$)	0–0.5 ($J_{1,4}$)
Furan	6.37, C-3	1.5–2 ($J_{2,3}$)	0.5–1 ($J_{2,4}$)	
	7.41, C-2	3–4 ($J_{3,4}$)	1.5–2 ($J_{2,5}$)	
Pyrrole	6.06, C-3	2–2.5 ($J_{2,3}$)	1.5 ($J_{2,4}$)	
	6.56, C-2	2.5–4 ($J_{3,4}$)	1.5–2 ($J_{2,5}$)	
Thiophene	7.00, C-3	4.5–6 ($J_{2,3}$)	1–2 ($J_{2,4}$)	
	7.18, C-2	3–4 ($J_{3,4}$)	2–3.5 ($J_{2,5}$)	
Pyridine	8.60, C-2	4.5–6 ($J_{2,3}$)	1–2.5 ($J_{2,4}$)	0.5–2 ($J_{2,5}$)
	7.25, C-3	7–8 ($J_{3,4}$)	0.5–2 ($J_{3,5}$)	
	7.65, C-4		0–0.5 ($J_{2,6}$)	

4.16.3.6 ^{13}C Chemical shift values for selected ring systems

Values of δ (ppm) are as follows:

Furan	142.6, C-2; 109.6, C-3.
Pyrrole	117.9, C-2; 107.8, C-3.
Thiophene	125.0, C-2; 126.4, C-3.
Pyridine	150.1, C-2; 123.9, C-3; 135.8, C-4.
Indole	124.6, C-2; 102.4, C-3.
Naphthalene	127.8, C-1; 125.8, C-2; 133.5 bridgehead.
Cyclopropane	−2.8.
Oxirane	40.0.
Aziridine	39.2.

4.16.3.7 $^{13}C-H$ coupling constants

A small number of representative $^{13}C-H$ coupling constants (Hz) are collected in Table 4.24.

No serious reference has been made to these coupling constants in the text. This is not a reflection on their usefulness in structural and stereochemical assignments but on their accessibility. In order to obtain these J values longer accumulation times are needed (the nOe enhancement is absent) and high-field instrumentation is required for improved dispersion and resolution. Hence they are expensive to determine. Two examples give an indication of their value in structural assignment.

Table 4.24. $^{13}C-H$ coupling constants

Compound	^1J	^2J	^3J	^4J
Butane	125*	5	2–8†	0
Ethene	155	2.5		
Ethyne	250	50		
Benzene	158	22	7	1
$CH_3{}^{13}CHO$	172	26		
$CH_3{}^{13}CH_2OH$	140	5		
$(CH_2O)_3$	166 (eq); 158 (ax)			

*In alkanes the J value decreases with increasing substitution: $CH_3, J = 125$; $CH_2, J = 120$; and $CH, J = 115\,Hz$.
†As with ^3J proton–proton coupling a Karplus equation relates the magnitude of J to the dihedral angle. Here an *anti* arrangement (angle = 180°) of H—C—C—^{13}C gives $J = 8\,Hz$, whereas the *gauche* configuration corresponds to $J = 2\,Hz$.

Example 1:

(4.32)   (4.33)   (4.32a)   (4.33a)

Surprisingly, neither infrared nor proton or carbon NMR spectroscopy distinguishes satisfactorily between **4.32** and **4.33** as the most likely product from a given reaction. The parent lactones **4.32a** and **4.33a** show a ^3J coupling for ^{1}H-3—^{13}C=O of 10 and 4 Hz, respectively, and a ^2J coupling for ^{1}H-2—^{13}C=O of about 4 and 0 Hz, respectively. The low-field carbonyl carbon of the reaction product had the appearance of a doublet of doublets with coupling constants of 4 and 12 Hz. Thus the structure **4.32** is confirmed.

Example 2:

(4.34)   (4.35)

The two products of another reaction might have had the structures **4.34** and **4.35**. The undecoupled ^{13}C spectrum showed both methyl groups as quartets (140 Hz), but further split differently (5.5 and 4.5 Hz) into doublets, the latter couplings being typical ^{13}C—^{1}H ^3J couplings through single bonds consistent with structure **4.34**. This excluded **4.35**, as the ^{13}C—^{1}H ^4J is zero and indicated the two diastereoisomers of **4.34** as products.

4.16.4 The TMS signal

The reasons stated for the use of TMS as an internal standard include the fact that it gives a sharp, singlet absorption. This is far from the truth. It might have been stated that *at 60 MHz* and *at low concentration* TMS gives what *appears to be* a sharp singlet. At 270 MHz it gives a spectrum containing five major lines, which are illustrated in Fig. 4.45a and b. The strong, central line in each is due to the protons in ^{28}Si(^{12}C^{1}H$_3$)$_4$, the doublet ($J = 6.5$ Hz) in the expanded spectrum (b)

NUCLEAR MAGNETIC RESONANCE SPECTROSCOPY

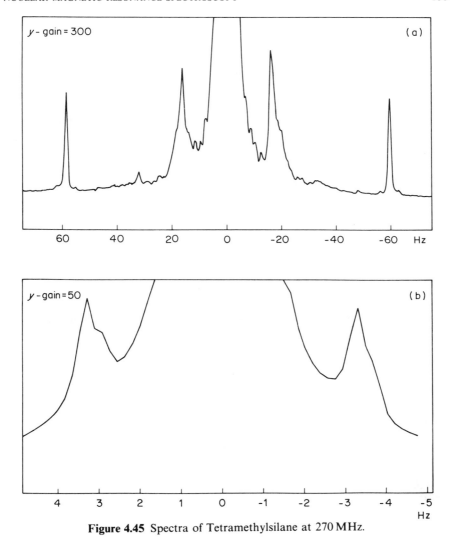

Figure 4.45 Spectra of Tetramethylsilane at 270 MHz.

to the *ca* 5% of $^{29}\text{Si}(^{12}\text{C}^1\text{H}_3)_4$ and the doublet ($J = 117\,\text{Hz}$) in the spectrum (a) corresponds to the *ca* 1% of $^{13}\text{CH}_3{}^{28}\text{Si}(^{12}\text{C}^1\text{H}_3)_3$ contained in natural-abundance samples of TMS. The apparent doublet, 'J' = 40 Hz, in (a) are sidebands from the main TMS signal and in practice are readily recognized as such. Even this is not the whole story, because there are finite numbers of TMS molecules having different combinations of ^{12}C, ^{13}C, ^{28}Si, ^{29}Si, ^1H, ^2H and ^3H, each having characteristic spectra but being present in concentrations beyond the sensitivity of current instrumentation.

4.16.5 Solvents

The ideal solvent would be cheap, unreactive, have a wide temperature range, say -150 to $+200\,°C$, over which it would remain mobile, contain no hydrogens and have good solvating properties for all kinds of molecules. Needless to say, no such ideal could exist.

The first choice is CCl_4 on account of its cheapness, but $CDCl_3$ is much the most commonly used, being a more powerful solvent. When the latter is used, the small trace of $CHCl_3$ present can be detected in proton spectra as a singlet absorption at 7.27 ppm. Another cheap solvent is hexachloroacetone. This is a better solvent than CCl_4 but it is difficult to recover the sample from it, as unfortunately is also the case with DMSO-d_6. The last compound is a very powerful solvent for many classes of compounds, including sugars and peptides, and can be used up to 180 °C.

Most of the common solvents in their fully deuterated form are employed as NMR solvents, including CD_2Cl_2, acetone-d_6 and methanol-d_4, all three capable of use down to near $-100\,°C$.

In FT measurements the use of some deuterated solvent in the sample is essential since the instrument is generally 'locked' on to the deuterium nuclei, that is, any potential drift in δ values of nuclei under examination is automatically corrected by reference to a chosen, usually the only, deuterium atom type in the solvent.

BIBLIOGRAPHY

Pretch, E., Clerc, T., Seibl, J. and Simon, W. (1983), *Spectral Data for Structure Determination of Organic Compounds*, Springer-Verlag, Berlin, Heidelberg, New York and Tokyo. A highly recommended collection of spectroscopic data for post graduate workers in organic chemistry.

Fuchs, P. and Bunnel, C. (1979), *Carbon-13 Based Organic Spectral Problems*, Wiley, New York and London. Provides a very useful collection of spectral problems.

Akitt, J. W. (1983), *NMR and Chemistry: An Introduction to the Fourier Transform Multinuclear Era*, 2nd ed, Chapman and Hall, New York and London.

Abraham, R. J. and Loftus, P. (1983), *Proton and Carbon-13 NMR Spectroscopy, an Integrated Approach*, Wiley-Heyden, London and New York. This book and that by Akitt are good texts to extend the subject matter covered by this book.

Pouchert, C. J. (1986), *The Aldrich Library of NMR Spectra*, 2nd ed, Aldrich Chemical Company, Milwaukee.

Bremser, W., Ernst, L., Franke, B., Gerharts, R. and Hardt, A. (1981), *Carbon-13 NMR Spectral Data*, Verlag-Chemie, Weinheim (mircrofiche).

Brügel, W. (1979), *Handbook of (^1H NMR) Spectral Parameters*, Heyden, London. This book and the previous two are excellent compilations of spectra or of spectral data.

Sanders, J. K. M. and Hunter, B. K. (1987), *Modern NMR Spectroscopy*, Oxford University Press, Oxford.

Derome, A. E. (1987), *Modern NMR Techniques for Chemistry Research*, Pergamon Press, Oxford. The last two books give an introduction to recent developments in 2D NMR techniques.

CHAPTER 5

Mass Spectrometry

5.1 INTRODUCTION

The previous chapters have considered the main spectroscopic techniques, involving the absorption of electromagnetic radiation, which can be used to aid in the elucidation of organic structures. Since about 1960, a further physical method, mass spectrometry, has increasingly been used to complement these spectroscopic methods. This method involves the separation and measurement of ions according to their mass-to-charge ratio. The popularity of the method is not hard to understand, since both molecular weight and molecular formulae can readily be determined with the expenditure of only minute amounts of sample. Further, the production of fragment ions leads to useful information concerning the structure of the parent molecule. As instrument design and technique have advanced, it has become possible to obtain mass spectra for most organic compounds including, in recent years, even thermally unstable and involatile species such as peptides with molecular weights of 10 000 or more. This chapter considers the interpretation of mass spectral data and its application to the determination of organic structures, and thus forms an essential complement to the main subject of this book, organic spectroscopy.

5.2 PRODUCTION OF SPECTRA

5.2.1 Inlet systems

5.2.1.1 Heated inlet

Volatile liquids may be injected through a septum into a small heated reservoir at a pressure of ca 10^{-2} Torr, which can be connected via a fine glass leak to the ion source at ca 10^{-6} Torr.

5.2.1.2 Direct inlet

Less volatile liquids and solids are generally introduced into the ion source in a glass smaple cup at the end of a metal probe shaft, after passing through a vacuum lock. The sample will then volatilize owing to the heat of the ion source. However, more control over the concentration of sample vapour can be achieved by using a probe, which allows the sample to be heated or cooled directly.

5.2.1.3 Gas chromatography

The effluent from capillary gas chromatographic columns consisting of a carrier gas, commonly helium, and sample vapour, can be passed directly into the ion source. This is possible because the source pumps can accommodate the relatively small gas flow-rate ($1-2\,cm^3\,min^{-1}$) from this type of column without too great an increase in the source pressure. The separation that may be achieved by the chromatography obviously makes this a most powerful method of analysing minute amounts of mixtures of compounds.

5.2.1.4 Liquid chromatography

The effluent from a high-performance liquid chromatograph consists of the eluting solvent plus the sample compound. This cannot be introduced directly into the ion source without too great an increase in pressure. Several methods have been developed for the removal of the eluting solvent, of which the recently developed thermospray method is possibly the most effective. The effluent is atomized and heated in a vacuum, allowing the less volatile sample to enter the ion source in the form of finely divided particles.

5.2.2 Ionization

The most commonly used method is *electron ionization* (EI). In this method, the sample, at a pressure of *ca* 10^{-6} Torr, is bombarded by a beam of electrons. The electrons are produced from a heated filament (often made of tungsten) and then accelerated across the ion source chamber to an anode held at a voltage difference that can be varied between 0 and 100 V. An energy of 70 eV is routinely used for ionization, since this has been found by experience to give a reproducible and maximized ion current. At this voltage numerous fragments are usually produced which are useful in structural elucidation. The lowest voltage at which ions are formed is called the *ionization potential*, and is generally around 10–12 eV. The bombardment produces both positive and negative ions. Although both types can be readily analysed, most EI spectra are concerned with the more abundant positive ions.

Other methods of ionization are important. The techniques of *chemical ionization* (CI), *field desorption* (FD) and *fast atom bombardment* (FAB), sometimes referred to as 'soft ionization' procedures (see Section 5.3.2 for details) produce spectra with fewer fragment ions than those given by EI. As a result, these methods often reveal the intact molecular ion, or a related protonated species, when this is absent from the EI spectra.

5.2.3 Analysis of ions

The ions produced in the ion source by the above processes may be separated according to their mass-to-charge ratio (*m*/*z*) using magnetic and/or electric fields. In a magnetic sector *single-focusing mass spectrometer*, the charged particles are first accelerated through a large potential (*V*), before being fired into a magnetic field (*B*); they then describe a radial path (radius *r*) governed by the equation

$$\frac{m}{z} = \frac{B^2 r^2}{2V} \tag{5.1}$$

Commonly the ions fall on a collector slit, before reaching a detector (Fig. 5.1). The width of this collector slit is variable, permitting changes in resolution (see Section 5.2.4). The most widely used detector is an electron multiplier, which can produce greater than a 2×10^6 gain in the ion current. These highly sensitive collectors further increase the inherent discrimination of the mass spectrometer against higher *m*/*z* ions. The enhanced signal is then amplified before being recorded (see Section 5.2.5).

A spectrum may be scanned by varying either the magnetic field (*B*) or the accelerating voltage (*V*). The *voltage scan* method has the advantage that very rapid scan rates are possible, but it is not often used because it limits the *m*/*z* range that can be examined for any fixed value of the magnetic field. The use of *magnetic scanning* is usual in modern mass spectrometers. It is possible to obtain a linear *m*/*z* scan by means of a 'Hall effect' probe, which senses the magnetic field, allowing precise control of its rate of change to be achieved.

The separation of ions by two alternative methods is important for the analysis

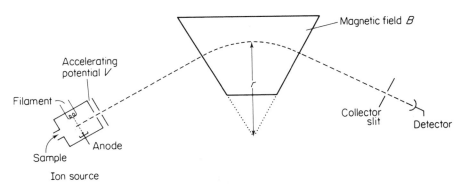

Figure 5.1 Single-focusing mass spectrometer.

of organic molecules, especially by combined gas chromatography–mass spectrometry (GC–MS). The first of these, the *quadrupole mass spectrometer*, separates ions by an electronic filtering action. The ions pass between four electrodes in the form of rods; at the same time they are subjected to a radio-frequency field such that only one particular m/z ion may pass the filter. Fluctuations of the radio-frequency field then permit extremely rapid scanning of the mass spectrum, which is perhaps the major advantage of the method. These spectrometers are also relatively compact, robust and less expensive, although of limited resolving power when compared with magnetic sector instruments. The second type, the *ion-trap mass-selective detector*, has the advantage of high sensitivity, when used in conjunction with a gas chromatograph. This is achieved by holding the ions in a cavity for a few microseconds, and thus increasing their abundance, before expelling them sequentially to yield the spectrum.

5.2.4 Resolution

The simplest method of defining resolution in a mass spectrometer is to consider two peaks of equal intensity, which are just resolved with a valley between them of an arbitrarily chosen 10% of their height. Then if these peaks have m/z values of x and $x + \Delta x$, the resolving power of the spectrometer is said to be 1 part in $x/\Delta x$ on

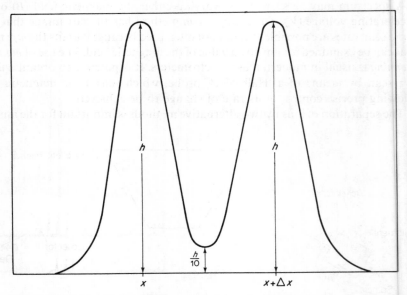

Figure 5.2 10% valley resolution. Reproduced with permission from S. F. Dyke, A. J. Floyd, M. Sainsbury and R. S. Theobald, *Organic Spectroscopy*, 2nd ed., Longman, London, 1978.

MASS SPECTROMETRY

a 10% valley definition (see Fig. 5.2), where Δx is the smallest increment in the m/z ratio x which can be so resolved.

Various factors influence the resolving power of a mass spectrometer: (a) the radius of the ion path, (b) the accelerating potential and strength of the magnetic field, (c) the width of source and collector slits, the minimum value of which is limited by detector sensitivity, and (d) the uniformity of kinetic energy for ions of the same m/z ratio passing through the magnetic analyser. It is the last factor which limits the resolving power of the *single-focusing* type of spectrometer to around 1 in 7500 on a 10% valley definition. In *double-focusing* instruments (see Fig. 5.7, Section 5.4.1) the ions are first focused by means of a radial electrostatic field before passing through the magnetic analyser. This results in an ion beam of uniform kinetic energy, which allows resolving powers of up to 1 in 150 000 to be attained.

5.2.5 Recording spectra

Recorders suitable for recording mass spectra require (a) very fast response times, typically up to 300 peaks per second, and (b) a high dynamic range because of the large variation ($> 10^3$) in the relative intensities of peaks in a spectrum. Until recently, these problems have resulted in the use of the inconvenient oscillograph,

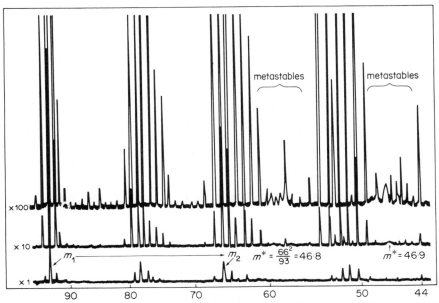

Figure 5.3 Oscillograph trace. Reproduced with permission from S. F. Dyke, A. J. Floyd, M. Sainsbury and R. S. Theobald, *Organic Spectroscopy*, 2nd ed., Longman, London, 1978.

140   ORGANIC SPECTROSCOPY

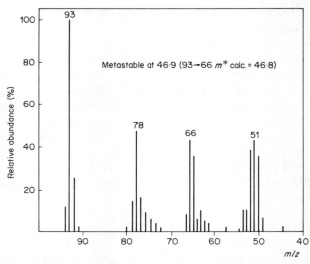

Figure 5.4 Bar graph. Reproduced with permission from
S. F. Dyke, A. J. Floyd, M. Sainsbury and R. S. Theobald,
Organic Spectroscopy, 2nd ed., Longman, London, 1978.

in which a series of mirror galvanometers of varying sensitivity reflect a beam of ultraviolet radiation into a moving strip of photographic paper, to record the spectra. The resultant series of traces (see Fig. 5.3) were difficult to comprehend readily, required laborious counting in order to assign m/z values and the photographic paper faded on exposure to daylight, unless chemically fixed. It is not surprising that modern mass spectrometers employ computers to collect and process the data, which can then be readily displayed after normalization as either a tabulation or bar-graph presentation (see Fig. 5.4). The visual impact of the bar-graph makes it especially suitable for interpretation purposes. The use of computers means that weak broad peaks, such as the metastable peaks observed in oscillograph traces (see Fig. 5.3), are no longer recorded. Similar information can, however, be obtained by alternative means (see Section 5.4.2).

5.3 DETERMINATION OF MOLECULAR FORMULAE

A mass spectrometer will readily provide the molecular weight of organic molecules, using only minute amounts of material. As a result, even a moderately accurate elemental analysis will then allow the molecular formula to be calculated.

5.3.1 The molecular ion

The removal of one electron from the whole molecule results in a species $M^{+\cdot}$, generally called the *molecular ion*. Thus the m/z value of the molecular ion gives the molecular weight of the sample.

$$M^{..} + e \rightarrow M^{+\cdot} + 2e$$

When examining the mass spectrum of an unknown compound, care must be taken when assigning the molecular ion. Firstly, the presence of natural isotopes means that most ions are represented by an 'isotopic cluster' reflecting the natural abundances of their constituent elements (see Section 5.3.3). Secondly, the highest m/z value cluster need not be due to the molecular ion, which may be too weak for certain detection. Again, the presence of higher molecular weight impurities may confuse the picture. Therefore, before selecting the molecular ion, as many checks as possible should be applied.

The validity of a candidate molecular ion should be checked, by considering whether the fragment ions in the spectrum could reasonably have been derived from it. Thus the loss of between 5 and 14 mass units is virtually impossible for most organic molecules, as it requires multiple loss of hydrogen atoms. The loss of 14 mass units often suggests a homologous impurity, since direct loss of methylene as a high-energy carbene is rarely, if ever, observed. The loss of 3 or 4 hydrogens can be observed; however, such loss is usually less intense than the loss of 1 or 2 hydrogens from the same molecular ion. Thus a specific loss of 3 hydrogens is extremely rare, so that two ions separated by 3 mass units are likely to arise, for example by loss of methyl and water from a molecular ion, which is not itself observed. Again, the loss of 19 or 20 mass units from a molecule which does not contain fluorine is an unlikely occurrence. In contrast, the loss of 15 mass units corresponding to a methyl radical is very common in any molecule possessing an alkane chain. Some of the common losses from the molecular ion are given later (see Table 5.3).

The stability of a molecular ion is, of course, related to its molecular structure. A knowledge of the chemical class to which an unknown belongs may, therefore, allow a prediction concerning likely molecular ion abundance. Thus, the presence of a conjugated π-electron system tends to give rise to an abundant molecular-ion peak. However, a structural feature that would stabilize a positive charge following fragmentation, especially with the loss of a stable neutral molecule, will tend to decrease molecular ion abundance. An approximate order for molecular ion abundance according to chemical class is aromatics and heteroaromatics > cycloalkanes, sulphides, thiols, conjugated alkenes > aldehydes, ketones, alkenes, carboxylic acids, amides, ethers, amines, n-alkanes > branched alkanes, halides, nitriles, alcohols, acetals.

It follows that a weak peak at the highest m/z ratio is unlikely to be the molecular ion for an aromatic compound. The presence of a higher molecular weight impurity should be considered. Likewise, an abundant ion at the highest m/z ratio in the spectra of a tertiary alcohol is unlikely to represent the molecular ion. Most probably it is related to the molecular ion by loss of water ($M - 18$).

Real molecular-ion peaks have an even m/z ratio, unless an odd number of nitrogen atoms is present in the parent molecule. This rule covers all organic molecules containing the common elements carbon, hydrogen, oxygen, nitrogen, phosphorus, sulphur, silicon and the halogens. It follows that no odd m/z peak can represent the molecular ion of a molecule which does not contain a nitrogen atom, but corresponds to a fragment ion, a nitrogenous impurity or, in rare cases, an ion produced by an ion–molecule reaction. Collisions between ions and molecules are uncommon at the low pressures used in the ion source. Nevertheless, for a very few samples, such collisions do lead to reaction, generally the abstraction of hydrogen by the molecular ion leading to an m/z $M + 1$ peak in the spectrum. The chemical ionization method (see Section 5.3.2), in which such collisions are encouraged, should confirm the true molecular weight of such a sample.

5.3.2 Soft ionization methods

Electron impact (EI) at 70 eV produces molecular ions with excess internal energy, which leads to various fragmentation processes. When the molecular ion has a fragmentation pathway of low activation energy available, then it will be short-lived and may fragment completely before leaving the ion source (ca 10^{-5}–10^{-6} s). In such cases no molecular-ion peak will be present in the mass spectrum (see Section 5.3.1). As the energy of the ionizing electrons is reduced towards the ionization potential (ca 10–12 eV) fragmentation is decreased, so that molecular ions are enhanced relative to fragment ion intensities. This method, involving the use of a *low ionization energy*, sometimes helps to identify weak molecular ions or indicate the presence of impurities. However, molecular ions which are absent at 70 eV are not revealed by reducing the ionizing potential. Alternative methods of ionization have been developed, which lead to molecular ions with a smaller excess of internal energy and often also of more stable form. These 'soft ionization' methods can frequently provide molecular ion information when electron bombardment fails.

5.3.2.1 Chemical ionization (CI)

The modern EI ion source can usually also provide *chemical ionization* as an alternative ionization mode. The changeover between EI and CI is rapid, in fact for gas chromatographic operation it is possible to arrange for automatic switching between scans only 1s apart, so that both EI and CI information can be obtained during one chromatagraphic run. The CI method involves mixing the sample (at ca 10^{-4} Torr) with a reactant gas at higher pressure (1 Torr) in the ion source. Commonly used reactant gases are methane, isobutane and ammonia. The resultant mixture is then subject to electron bombardment, which intially ionizes some reactant gas molecules. Thus methane gives rise to the expected $CH_4^{+\cdot}$ and CH_3^{+} species. At the high source pressures used, collisions and ion–

molecule reactions within the reactant gas are common, leading to secondary ions with a small excess of internal energy:

$$CH_4^{+\cdot} + CH_4 \rightarrow CH_5^+ + CH_3^{\cdot}$$
$$CH_3^+ + CH_4 \rightarrow C_2H_5^+ + H_2 \qquad (5.2)$$
Secondary ions

These secondary ions eventually collide with sample molecules, resulting in ionization of the latter. Ionization is commonly due to protonation (equation 5.3), especially for basic (even moderately basic, e.g. alcohols) compounds.

$$M + CH_5^+ \rightarrow (M+1)^+ + CH_4 \qquad (5.3)$$

Further, these *quasi-molecular ions* are even-electron species, which tend to be more stable than the radical molecular ions produced by electron impact. A combination of lower energy in the ionization process with this greater stability means that the quasi-molecular ions are usually fairly abundant in CI spectra.

Interpretation of these spectra is often more straightforward than for EI spectra, since there are generally fewer fragment ions, which tend to be of greater significance. The amount of fragmentation can be changed by the nature of the reactant gas. In general, both the range of compounds protonated and the degree of fragmentation observed decreases as the reactant gas is changed in the order methane > isobutane > ammonia. In fact, ammonia will only protonate fairly basic molecules such as alcohols and amines.

5.3.2.2 Fast atom bombardment (FAB)

Although the CI method greatly increases the range of samples for which molecular ion information can be obtained, it still requires molecules that are moderately volatile and thermally stable. This excludes many compounds of biological importance, such as peptides and carbohydrates. The recently developed technique of *fast atom bombardment* (FAB) overcomes these problems. Fast atoms are obtained from an ion gun, by allowing the accelerated ion beam to enter a collision chamber and exchange energy with neutral atoms (equation 5.4):

$$Xe^+ \text{(fast)} + Xe \rightarrow Xe^+ + Xe\text{(fast)} \qquad (5.4)$$

Argon is often used, but the greater mass of xenon means that the fast atom beam has higher energy and results in the production of more ions from the sample. The resultant fast atoms are directed on to the sample, held in a liquid matrix on a metal target (see Fig. 5.5). The liquid matrix is most commonly glycerol. The metal target is held at a potential difference with respect to the source slits, so that ions formed, at or just above the surface of the sample on impact of the fast atoms, are expelled from the source. By reversing the nature of the potential, either

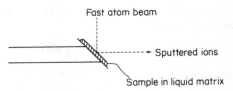

Figure 5.5 Fast atom bombardment.

positively or negatively charged ions may be examined. The method is extremely simple and gives rise to spectra at near room temperature. The method is most effective for relatively polar substances which will readily either accept or give up a proton. Generally, abundant quasi-molecular ions [i.e. $(M + 1)^+$ or $(M - 1)^-$] plus a few diagnostic fragment ions result. A feature of FAB spectra that is often observed is the presence of glycerol adducts or molecular clusters as weak peaks at higher m/z values [i.e. $(M + G + 1)^+$, where $G = 92$ for glycerol, or $(2M + 1)^+$ ions]. The method has proved extremely valuable for polar molecules which are relatively involatile and thermally unstable. Hence the mass spectra of molecules such as peptides and glycosides can be obtained without prior derivatization.

5.3.3 Accurate mass measurement

Atomic weights of atoms are non-integral (see Table 5.1) when compared with carbon-12 as a standard (12.0000). As a result, two molecules or fragments may

Table 5.1. Natural abundance and atomic weights of some common isotopes

Isotope	Natural abundance	Atomic weight
^{1}H	99.98	1.0078
^{2}H	0.02	2.0141
^{12}C	98.93	12.0000
^{13}C	1.07	13.0034
^{14}N	99.63	14.0031
^{15}N	0.37	15.0001
^{16}O	99.76	15.9949
^{17}O	0.04	16.9991
^{18}O	0.20	17.9992
^{19}F	100.00	18.9984
^{28}Si	92.17	27.9769
^{29}Si	4.71	28.9765
^{30}Si	3.12	29.9738
^{31}P	100.00	30.9738
^{32}S	95.03	31.9721
^{33}S	0.75	32.9715
^{34}S	4.22	33.9679
^{35}Cl	75.53	34.9689
^{37}Cl	24.47	36.9659
^{79}Br	50.52	78.9183
^{81}Br	49.48	80.9163
^{127}I	100.00	126.9044

have the same integral mass, although differing in their non-integral masses. Thus the ions A and B of nominal mass 200:

$$\begin{array}{lll} \text{A} & C_{11}H_{24}N_2O & 200.1883 \\ \text{B} & C_{10}H_{20}N_2O_2 & 200.1520 \\ & & \overline{\Delta x = 0.0363} \end{array}$$

differ in mass by 0.0363 and thus require a resolution about 1 in 5500 to separate them, which is readily achieved by a double-focusing mass spectrometer (see Section 5.2.4). Further, the m/z value of such an ion can be measured with a precision of 1 or 2 parts in 10^6 by comparison with an ion of precisely known m/z ratio. Thus the molecular weight of A or B could be found with an error of ± 0.0004, which in most cases would allow the molecular formulae to be decided between only a few possibilities. In general, computer analysis is the best method for selecting suitable formulae to fit the observed molecular weight, although at times the tabulations of Beynon and Williams (1963) or the computational method of Lederberg (1964) may be found useful.

5.3.4 Natural abundance of isotopes

In considering molecular ions, account must be taken of the fact that many elements consist of two or more isotopes (see Table 5.1). The molecular ion and its associated molecular weight M is considered to be derived from the most abundant isotopes of the elements present, which for all the common elements constitute the lightest isotopes. Of course, the mass spectrometer records not only such ions, but also ions containing one or more of the heavy isotopes of its constituent atoms. It is important to realize, therefore, that almost all ions consist of a cluster of peaks, with intensity ratios characteristic of their elemental composition. In the case of the *molecular-ion cluster* at m/z $M, M + 1, M + 2$, etc., this pattern can be used to assign the molecular formulae. However, the molecular ion needs to be fairly strong and special care taken to obtain accurate intensity data before this becomes a reliable method. Generally, accurate mass measurement is a more certain method for the determination of the elemental composition of an ion. Nevertheless, isotopic ratios are an easy method for determining the number of chlorine or bromine atoms present, detecting the presence of sulphur and making a quick assessment of the probable number of carbon atoms in a molecule.

5.3.4.1 Carbon, hydrogen, oxygen and nitrogen isotopes

The isotopic abundances for these elements is fairly low, so that the $M + 1$ and $M + 2$ peaks are small unless large numbers of these atoms are present. Thus, for a single carbon atom the ratio $[(M + 1)/M] \times 100 = 1.07/98.93 \times 100 = 1.08\%$ due to the natural abundance of ^{13}C atoms. For ions containing n carbon atoms, the probability arises that there will be 1.08 n% of these ions with a ^{13}C atom present, leading to an $M + 1$ peak with that intensity relative to the peak M. Similar allowance can be made for the isotopes of hydrogen, nitrogen and oxygen. The effect of oxygen is more pronounced on the $(M + 2)/M$ ratio than on the $(M + 1)/M$, owing to the greater abundance of ^{18}O rather than ^{17}O isotopes. The $(M + 2)/M$ ratios are generally small, since for carbon, hydrogen and nitrogen isotopes, two heavy isotopes of low abundance must be present in the same ion. The tabulations of Beynon and Williams (1963) list $(M + 1)/M$ and $(M + 1)/M$ ratios for all possible combinations of C, H, N and O atoms up to a molecular weight of 500.

5.3.4.2 Sulphur isotopes

The relatively large $(M + 2)/M$ ratio of 4.44% due to the natural abundance of ^{34}S isotopes makes the presence of even one sulphur atom in an ion readily detected. It should be realized that in the Beynon and Williams (1963) compilation for C, H, N and O combinations up to molecular weight 500, the $(M + 2)/M$ ratios never exceed 10%. To illustrate the point, ratios for some ions of nominal $m/z = 100$ are given in Table 5.2.

Table 5.2. Some isotopic ratios

Elemental composition	$(M + 1)/M$ (%)	$(M + 2)/M$ (%)
$C_5H_{12}N_2$	6.38	0.20
C_5H_8S	6.34	4.46
$C_5H_8O_2$	5.60	0.56
$C_5H_2F_2$	5.44	0.15

5.3.4.3 Halogen isotopes

The monoisotopic fluorine and iodine atoms do not contribute to isotopic patterns, but they may still be recognized by the lower than expected $(M + 1)/M$ and $(M + 2)/M$ ratios based on C, H, N and O alone for the required molecular weight. This is especially true for ^{127}I atoms, which would replace a considerable number of such atoms.

Both chlorine (^{35}Cl and ^{37}Cl, ratio 3:1) and bromine (^{79}Br and ^{81}Br, ratio 1:1) have large natural abundances of their heavy isotopes, which result in very characteristic patterns for ions containing them. For molecular ions, the peaks at $M, M + 2, M + 4$, etc., are affected, while the $(M + 1)/M$ ratio still reflects the C, H, N and O composition being unaffected by chlorine or bromine isotopes. The intensity ratios for halogen-containing molecular ion clusters may be estimated using the binomial expansion $(a + b)^n$, where a is the abundance of the

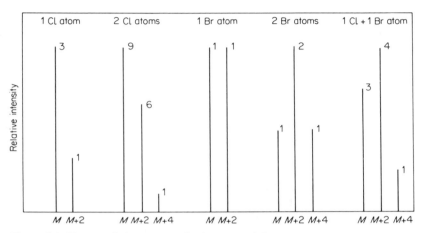

Figure 5.6 Characteristics patterns for ions containing one or two chlorine and/or bromine atoms

lighter isotope, b is the abundance of the heavier isotope and n is the number of halogen atoms. If $n = 3$, then $a^3 + 3a^2b + 3ab^2 + b^3$ is the binomial expansion.

It can be seen that three halogen atoms will give rise to four peaks, one for each term in the expansion. In fact, n halogen atoms will always produce a molecular-ion cluster of $n + 1$ peaks, each separated by two units. Terms in the expansion with only a present correspond to the $M^{+\cdot}$ peak, as each a is replaced by b two mass units are added. Thus for chlorine, if $n = 3$ then $a = 3$ and $b = 1$, and

$$a^3 + 3a^2b + 3ab^2 + b^3 = 27 + 27 + 9 + 1$$

which corresponds to a molecular-ion cluster M, $M + 2$, $M + 4$, $M + 6$ with intensity ratios 27:27:9:1. For bromine, if $n = 3$ then $a = 1$ and $b = 1$, and

$$a^3 + 3a^2b + 3ab^2 + b^3 = 1 + 3 + 3 + 1$$

which corresponds with a molecular-ion cluster M, $M + 2$, $M + 4$, $M + 6$ with intensity ratios 1:3:3:1. Mixtures of chlorine and bromine can be dealt with in a similar manner by using the product of two binomial expansions $(a + b)^n(c + d)^m$, where n and m are the number of chlorine and bromine atoms present, respectively. The characteristic patterns observed for ions containing either one or two halogen atoms are illustrated in Fig. 5.6.

5.4 FRAGMENTATION PROCESSES

The minimum energy needed to produce ionization of a molecule is termed the *ionization potential*. In the electron bombardment method this energy usually corresponds to between 10 and 15 eV for the electron beam. When the ionization is obtained under the normal conditions of a 70 eV beam of electrons, some electrons interact weakly to produce molecular ions with little excess energy. In some cases, however, there is a greater transfer of energy leading to excited molecular ions, which in turn fragment, generally by intramolecular processes. Molecular collisions, leading to energy interchange or reaction between species, are a rare event at the relatively low pressure (ca 10^{-6} Torr) of the ion source. Molecular ions produced with insufficient energy to undergo fragmentation are thus observed as such, despite the high energy of the ionizing beam. Of course, when there is an intramolecular fragmentation process of very low or zero activation energy available, then no molecular ion appears in the mass spectrum, and further, even reducing the electron-beam energy to near the ionization potential will fail to produce one.

The fragmentations themselves are not random, since the excitation energy is best relieved when the bonds broken lead to well stabilized fragments. In consequence, the most abundant fragment ions may be related to the molecular structure of the sample, using the same principles that chemists use to predict reaction pathways. The analogy is closest to high-energy processes such as pyrolysis in the gas phase, rather than the more commonly encountered solution chemistry. The primary fragment ions often have sufficient energy to fragment further, so leading to the complete spectrum.

5.4.1 Metastable ions

The fragmentation of an ion (parent) will yield a new ion (daughter) plus either a neutral molecule or a radical (equation 5.5).

$$m_1^+ \rightarrow m_2^+ + \text{non-charged particle} \tag{5.5}$$

Normally, both the parent m_1^+ and the daughter m_2^+ ions are observed in the mass spectrum, although there is no direct evidence that m_2^+ was derived from m_1^+ in a one-step process. Thus the ions m_1 and m_2 which are present in the ion source,

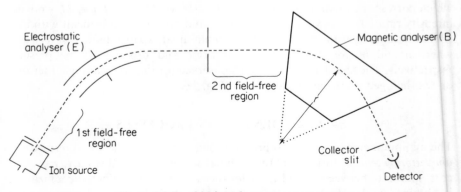

Figure 5.7 Double-focusing mass spectrometer.

MASS SPECTROMETRY

accelerated and collected according to the equation $m/z = B^2r^2/2V$ (see Section 5.2.3), are called normal ions. In general, ions which decompose within the electrostatic or magnetic analyser are not focused. However, ions which decompose within the second field-free region of a double-focusing mass spectrometer (Fig. 5.7) result in an ion m_2^+ with less kinetic energy than a normal ion.

These ions are called *metastable ions* and are focused as broad, low-abundance peaks (see Fig. 5.3, Section 5.2.5) at a different m/z value to a normal m_2^+ ion. There is a simple relationship connecting the apparent m/z value (m^*) of the metastable ion, with those of its parent (m_1) and daughter (m_2) ions (equation 5.6).

$$m^* = (m_2)^2/m_1 \tag{5.6}$$

Metastable peaks need not occur as integral m/z ratios. Since a metastable ion results from fragmentation in a very restricted time span, namely that required to traverse the second field-free region, it is usually safe to conclude that it can only result from a one-step process. Thus, the observation of a metastable peak m^* fitting the relationship of equation 5.6, provides evidence that m_1 and m_2 are related by a one-step process and derived from the same molecular species. However, absence of a suitable metastable peak m^* is not evidence that m_1 and m_2 are not so related. When correlating an observed metastable peak with its corresponding parent and daughter ions, use can be made of the compilations of Beynon *et al.* (1965). The typical appearance of a metastable peak in a galvanometer recording is shown in Fig. 5.3 (Section 5.2.5).

5.4.2 Linked scans

The wide use of computers to record mass spectra means that metastable ions are generally no longer recorded. However, the use of computer control has allowed linked scanning, in which the voltage (E) of the electrostatic sector may be varied along with the magnetic field (B) (see Fig. 5.7), so as to keep a fixed relationship between them while scanning. These methods yield similar information to metastable ions, although in a more definitive manner. When the ratio B/E for a particular parent ion m^+ is found and then B and E are scanned simultaneously so that B/E remains constant, then only metastable ions derived from m^+ are observed. Conversely, when the B^2/E value appropriate to a particular daughter ion is set up and this time B^2/E is held constant while B and E are scanned, then only parent ions are located.

5.4.3 Even-electron and odd-electron ions

Most organic compounds on electron bombardment yield a molecular-ion radical or odd-electron ion. The symbol $M^{+\cdot}$ is used to denote this radical nature, although it should be remembered that an electron is removed, not added, as the symbol may be thought to suggest. Loss of a neutral molecule from the molecular ion leads to an odd-electron fragment ion. All odd-electron ions have even-numbered m/z ratios, unless they contain an odd number of nitrogen atoms. Again, the loss of a neutral radical from an odd-electron ion yields an even-electron ion, which will be observed at an odd-numbered m/z ratio unless it contains an odd number of nitrogen atoms. Whereas odd-electron ions may fragment by loss of either a radical or an even-electron molecule, the more stable even-electron ions rarely lose other than an even-electron molecule. As a result, mass spectra generally contain more even-electron ions than odd-electron ions. The latter, where they do occur, are often of structural significance. Hence, the mass spectra of aliphatic compounds usually exhibit homologous series of even-electron ions, corresponding to carbon chains, and it is rare to find odd-electron ions in such an homologous series.

5.4.4 Symbols

In discussing fragmentation processes, the usual arrow and fish-hook symbols ⤴ and ⤵ will be used to represent the heterolytic and homolytic cleavage of bonds:

$$X-Y \longrightarrow X^+ + Y^- \qquad \text{Heterolytic cleavage}$$

$$X-Y \equiv X-Y \longrightarrow X^\cdot + Y^\cdot \qquad \text{Homolytic cleavage}$$

For odd-electron ions, the site from which the electron has been removed may not always be indicated. In these cases the fish-hook symbol is assumed to act at the electron-deficient bond:

$$[X-Y]^{+\cdot} \longrightarrow X^+ + Y^\cdot$$

5.4.5 Simple fission

This term describes either the homolysis of a bond in an odd-electron ion to yield an even-electron fragment ion plus a radical:

$$A-B^{+\cdot} \longrightarrow A^\cdot + B^+$$

or the heterolysis of a bond in an even-electron fragment ion to yield another even-electron ion plus a neutral molecule:

$$A-B^+ \longrightarrow A^+ + B$$

These simple fission processes lead to even-electron ions *via* one-bond cleavage. Naturally, the rate of these processes depends on the transition states involved. If these resemble the product ion, as often happens, then stable fragments tend to correspond to abundant peaks in the spectra. Since for alkyl carbonium ions the order of stabilities is tertiary > secondary > primary, it is not surprising that fragment abundances follow the same order. Thus the greater abundance of peaks corresponding to tertiary and secondary carbonium ions in a homologous series allows the position of any chain branching to be located. Again, the presence of a heteroatom (especially nitrogen) with a lone pair of electrons, adjacent to the carbonium ion, will result in a very favourable mesomeric stabilization (**5.1**). An adjacent π-electron system will similarly assist

(**5.1**)

the formation of a carbonium ion. In the spectra of alkylbenzenes (**5.2**), for example, a prominent peak is observed at m/z 91, which corresponds to the benzyl cation (**5.3**). However, deuteration studies have shown the true structure to be the

m/z 91

(**5.2**)   (**5.3**)   (**5.4**)

tropylium ion (**5.4**). This is not surprising, when it is realised that this is one of the more stable carbonium ions, having six π-electrons and aromatic character.

5.4.6 Elimination and rearrangement

These processes involve the breaking of at least two bonds while an atom or group of atoms is transferred from one site to another. The most common transfer involves an hydrogen atom:

$$A-B \longrightarrow A^{+\cdot} + BH \quad \text{elimination}$$

$$A-B \longrightarrow AH^{+\cdot} + B \quad \text{rearrangement}$$

Elimination and rearrangement processes generally involve odd-electron ions yielding an odd-electron daughter ion plus a neutral molecule. Elimination is characterized by hydrogen transfer to the eliminated neutral molecule, whereas in rearrangement the hydrogen is transferred to the odd-electron daughter ion.

As usual, the stabilities of the resultant daughter ions and neutral species are often related to the abundance of this type of ion in the spectra. Naturally, the loss of small stable molecules (BH and B) is frequently important. Molecules that are often eliminated include H_2O, RCO_2H, HCN, H_2S and HX (X = halogen), whereas rearrangement leads to the loss of species such as $RCH=CH_2$, $RC\equiv CH$, CH_2O, $RCH=C=O$, CO and CO_2. When elimination or rearrangement can occur *via* a six-membered cyclic transition state, then such processes are often very prominent (see Section 5.5.1.7). The occurrence of three-, four- and five-membered cyclic transition states has also been postulated; indeed, the requirement for intramolecular processes means that four-membered cyclic transition states are more commonly encountered than expected by comparison with reactions in solution.

5.5 FRAGMENTATIONS ASSOCIATED WITH FUNCTIONAL GROUPS

The mass spectra of monofunctional compounds will be related to (a) their carbon skeleton, e.g. branched or unbranched alkane, aromatic, and (b) the nature of the functional group present. In difunctional compounds, the relative abilities of the functional groups to control fragmentation will complicate prediction of the likely mass spectrum.

5.5.1 Aliphatic compounds

5.5.1.1 Alkanes

Hydrocarbon spectra are important in the sense that most functionalized aliphatic compounds show typical alkane fragments in their spectra, together with peaks characteristic of the functional groups present.

The molecular ion **5.5** is best considered as a non-localized structure, since the electron-deficient bond cannot be specified. The mass spectra of alkanes show a homologous series of even-electron ions, corresponding to m/z $C_nH_{2n+1}^{+\cdot}$. Primary fragmentation by simple fission leads to even-electron ions such as **5.6**, which can in turn undergo simple fission losing an alkene, with the formation of further ions **5.7** in the homologous series. These secondary fragmentations lead to a greater abundance of smaller fragments formed after two or three such processes. Indeed, n-alkanes generally exhibit maximum abundance around $C_3H_7^+$ or $C_4H_9^+$ (i.e. m/z 43 or 57). Associated with each $C_nH_{2n+1}^+$ peak are less intense peaks at m/z 1, 2 or 3 mass units lower caused by loss of hydrogen radicals. As might be expected, the even-electron ion series $C_nH_{2n-1}^+$ is often prominent. Naturally, isotopes give rise to peaks at m/z 1 and 2 mass units higher with intensities appropriate to their natural abundance. The presence of chain branching gives rise to secondary or tertiary carbonium ions by fission at the point of branching, which results in a more abundant peak in that region of the spectrum, usually accompanied by a lower molecular ion abundance.

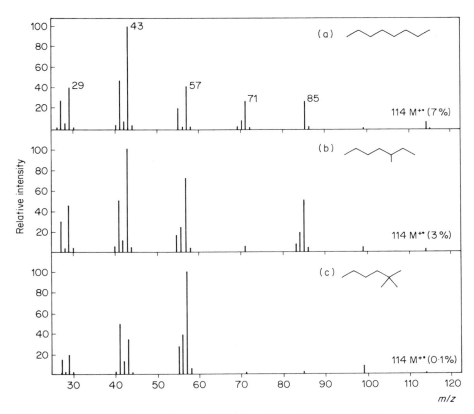

Figure 5.8 70 eV EI Spectra of some C_8 hydrocarbons: (a) octane; (b) 3-methylheptane; (c) 2,2-dimethylhexane.

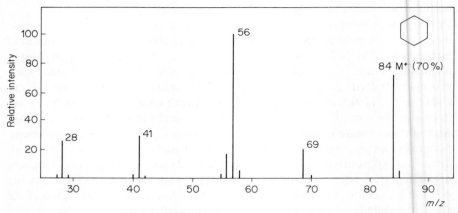

Figure 5.9 70 eV EI spectrum of cyclohexane.

The 70 eV EI spectra of some C_8 saturated hydrocarbons (Fig. 5.8) illustrate some of these points. The even-electron ion series $C_nH_{2n+1}^+$ can be observed in the spectra at m/z 29, 43, 57, 71, 85 and 99. The presence of $C_nH_{2n-1}^+$ ions, especially m/z 27 and 41, can also be observed. The effect of chain branching in decreasing the molecular-ion intensity is clear. The changes in intensity of the $C_nH_{2n+1}^+$ ions also indicates the chain branching. Thus in 3-methylheptane (Fig. 5.8b) increased abundance for m/z 57, 85 and 99 is observed, corresponding to the formation of secondary carbonium ions with loss of n-butyl, ethyl and methyl radicals, respectively. While, in the spectrum of 2,2-dimethylhexane (Fig. 5.8c) the branching is clearly shown by the formation of a *tert*-butyl carbonium ion at m/z 57.

5.5.1.2 Cycloalkanes

Breaking one bond in a cyclic system leaves the m/z ratio unchanged, so a second bond must be broken before a fragment peak is recorded in the mass spectrum. This results in a stronger molecular ion for cyclic compounds. Once the ring is cleaved elimination of an alkene is the favoured process. The presence of a saturated ring in a hydrocarbon chain is similar to chain branching in alkanes. The spectrum of cyclohexane (Fig. 5.9) illustrates a few of these features. The prominent loss of 28 mass units can be seen as cleavage of the ring (**5.8**), followed by loss of ethene.

5.5.1.3 Alkenes and Alkynes

Small alkenes and alkynes show stronger molecular ions than the alkanes; however, above C_6 this situation is gradually reversed. The even-electron ions belonging to series corresponding to $C_nH_{2n-1}^+$ for alkenes and $C_nH_{2n-3}^+$ for alkanes are generally prominent in the mass spectra. Under electron impact multiple bonds, especially in alkenes, tend to migrate. Consequently, alkene isomers, even those involving changes in the carbon skeleton, often show very similar mass spectra.

5.5.1.4 Alcohols, thiols and primary amines

The presence of a heteroatom, with lone pairs of electrons, means a lower ionization potential compared with the alkanes. Indeed, it is reasonable to assume that the molecular ion **5.9** will be of a more localized nature, in which one of the lone pair electrons has been lost. This will then lead to simple fission of

$$R-\ddot{X}H \xrightarrow{-e} R-\overset{+\cdot}{\ddot{X}H}$$
$$(X=O, S \text{ or } NH) \quad (\mathbf{5.9})$$

bonds adjacent to the heteroatom with the loss of an alkyl radical and the formation of even-electron ions (**5.10**), stabilized by mesomerism. The effectiveness of this stabilization decreases in the series $N > S > O$, so the peaks at $m/z = 30 + (n \times 14)$ are dominant in the spectra of amines, whereas for thiols and alcohols the $m/z = 47 + (n \times 14)$ and $m/z = 31 + (n \times 14)$ peaks, respectively, are relatively less abundant. Generally, the preferred loss of the alkyl radical $R^{1\cdot}$ will be controlled by the relative stabilities of the alternative radicals R^1, R^2 and R^3.

$$X = O \; m/z = 31 + (n \times 14)$$
$$X = NH \; m/z = 30 + (n \times 14)$$
$$X = S \; m/z = 47 + (n \times 14)$$

(**5.10**)

The fragment ions **5.10** can themselves undergo further degradation by rearrangement with the expulsion of an alkene, when R^2 or R^3 is an ethyl group or larger. The resultant ions **5.11** still in fact belong to the same homologous series **5.10**. The elimination of water from alcohols $(M - 18)$ and to a lesser extent

(5.10)   (5.11)

hydrogen sulphide from thiols ($M - 34$) is an important process, although the elimination of ammonia ($M - 17$) is not usually observed in the spectra of primary amines. These eliminations result in odd-electron ions with an alkene composition ($C_nH_{2n}^{+\cdot}$), which by simple fission processes can give rise to peaks in the series $m/z = C_nH_{2n-1}^+$, which are usually especially noticeable in the spectra of alcohols. The spectra of alcohols should be obtained at the lowest practicable temperature, since degradation and dehydration can occur thermally, in addition to under electron impact. The spectra in Fig. 5.10 illustrate some of these features.

The spectrum of butan-1-ol (Fig. 5.10a) shows a base peak at m/z 31 due to fission adjacent to the heteroatom, which suggests a primary alcohol. As is typical of alcohols, the molecular ion is fairly weak. Elimination of water yields a strong ion at m/z 56 ($C_4H_8^{+\cdot}$), accompanied by ions in the series $C_nH_{2n-1}^+$ (m/z 27, 41).

The tertiary 1,1-dimethylethanethiol spectrum (Fig. 5.10b) exhibits a stronger molecular ion (m/z 90), even when compared with the primary alcohol. The loss of a thiol radical ($M - 33$) results in a strong peak at m/z 57; it is common with secondary and tertiary thiols that the formation of a stable even-electron carbonium ion (e.g. **5.12**, m/z 57) is favoured over expulsion of hydrogen sulphide.

(5.12)

The expected cleavage adjacent to the heteroatom with loss of a methyl radical and formation of a resonance-stabilized ion (**5.10**, m/z 75; X=S, $R^2 = R^3 = CH_3$) is observed. The base peak at m/z 41 is surprisingly large in view of the need for structural rearrangement in its formation.

The spectrum of 2-butylamine (Fig. 5.10c) is dominated by cleavage adjacent to the nitrogen, leading to the very well stabilized ions **5.13** and **5.14**. Further, as would be expected, the loss of the more stable ethyl radical (process a) is favoured over the loss of a methyl radical (process b).

m/z 44

(5.13)

m/z 58

(5.14)

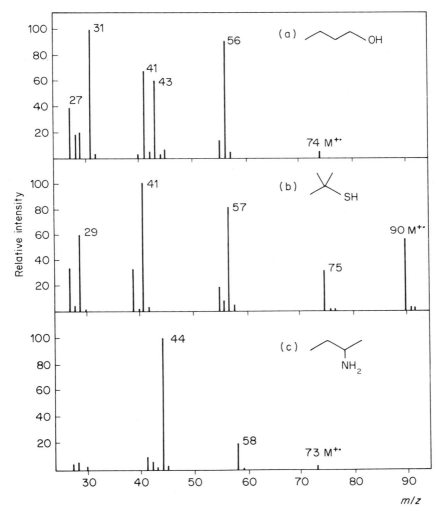

Figure 5.10 70 eV EI spectra of (a) butan-1-ol, (b) 1,1-dimethylethanethiol and (c) 2-butylamine.

5.5.1.5 Cyclic alcohols, thiols and amines.

It can be assumed that cleavage of the ring is triggered by ionization of the heteroatom, but further rearrangement and fission are required before any change in m/z value is observed. Ions of the type **5.15** result. Loss of water $(M - 18)$ is still characteristic of these alcohols, whereas cyclic thiols, being secondary, show losses of both hydrogen sulphide $(M - 34)$ and a thiol radical $(M - 33)$.

X = O m/z = 57
X = S m/z = 73
X = NH m/z = 56

(5.15)

The spectra in Fig. 5.11 illustrate most of these features. Thus the ions **5.15**, which result from cleavage adjacent to the heteroatom followed by rearrangement and simple fission, may be observed in all of the spectra. This is the dominant feature in the spectrum of the amine (Fig. 5.11c) at m/z 56 (**5.15**,

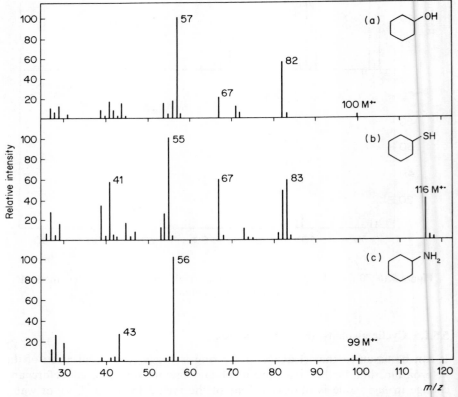

Figure 5.11 70 eV EI spectra of (a) cyclohexanol, (b) cyclohexanethiol and (c) cyclohexylamine.

X = NH), is important in the spectrum of the alcohol (Fig. 5.11a) at m/z 57 (**5.15**, X = O) and is least important with the thiol (Fig. 5.11b) at m/z 73 (**5.15**, X = S). Cyclohexanol shows ready loss of water ($M - 18$; m/z 82), whereas cyclohexanethiol loses hydrogen sulphide ($M - 34$; m/z 82) and a thiol radical ($M - 33$; m/z 83) equally. The presence of a metastable ion (m/z 54.7) suggests that the ions at m/z 67, observed in the spectrum of the alcohol (and the thiol), arise from the $[M - 18]^{+\cdot}$ ions (or $M - 34$ for thiol) at m/z 82.

5.5.1.6 Ethers, sulphides and secondary and tertiary amines

The spectra of these classes of compounds have similarities with those of alcohols, thiols and primary amines. As before, it is helpful to consider the molecular ions as being formed by removal of an electron from a heteroatom lone pair. This allows ready simple fission to occur adjacent to the heteroatom, with preferential loss of the most stable radical.

$$X = O, \quad m/z\ 45 + (n \times 14)$$
$$X = NR, \quad m/z\ 44 + (n \times 14)$$
$$X = S, \quad m/z\ 61 + (n \times 14)$$

(**5.16**)

The resonance-stabilized carbonium ions **5.16** have m/z values in homologous series. Further rearrangement of these ions is possible with loss of alkene molecules, leaving their m/z values in the same homologous series.

Elimination of alcohols from ethers is generally not observed. Instead, simple fission of the C—O bond is usually more important (equation 5.7). This process has already been described for thiols, where it occurs when the resultant carbonium ion is sufficiently stable (see Fig. 5.10b). The more stable of the

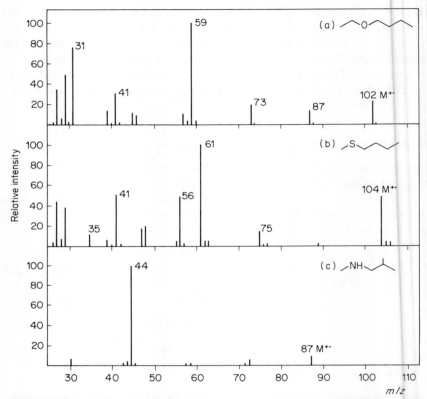

Figure 5.12 70 eV EI spectra of (a) ethyl butyl ether, (b) methyl butyl sulphide and (c) N-methylisobutylamine.

carbonium ions R^{1+} or R^{2+} is generally favoured. Amines rarely give rise to ions by C—N bond fission. The spectra in Fig. 5.12 illustrate some of these fragmentations.

$$R^1 \overset{a}{\underset{}{\diagdown}} \overset{+\cdot}{X} \overset{b}{\underset{}{\diagup}} R^2 \quad \begin{matrix} a \\ \longrightarrow \\ b \end{matrix} \quad \begin{matrix} R^{1+} + R^2X^\cdot \\ R^1X^\cdot + R^{2+} \end{matrix} \qquad (5.17)$$

The spectrum of the ether (Fig. 5.12a) shows a base peak (**5.17**, m/z 59) due to simple fission adjacent to the oxygen atom and loss of a propyl radical; the corresponding loss of a methyl radical to yield **5.18** (m/z 97) is much weaker.

Both **5.17** and **5.18** can undergo rearrangement with loss of alkene to yield **5.19** (m/z 31).

[Scheme showing fragmentations producing ions at m/z 59 (5.17), m/z 97 (5.18), m/z 31 (5.19), and structures 5.18, 5.17]

In the case of the sulphide spectra (Fig. 5.12b), again simple fission adjacent to the sulphur atom yields the most abundant ion **5.20** (m/z 61), whilst elimination of methanethiol ($M - 48$) leads to a fairly abundant peak at m/z 56. Again, the amine spectrum is totally dominated by cleavage adjacent to the nitrogen atom, to give the resonance stabilized ion **5.21** (m/z 44).

[Fragmentation schemes for 5.20 (m/z 61) and 5.21 (m/z 44)]

5.5.1.7 Aldehydes and ketones

Again, it is convenient to consider the molecular ion of these compounds to have the charge-localized structure **5.22**.

[Scheme showing cleavage of ketone 5.22 into R$^{1\cdot}$ + R^{2}C≡O$^+$ (5.23) and R^{1}C≡O$^+$ + R$^{2\cdot}$ (5.24)]

Simple fission adjacent to the carbonyl group will lead to either of the acylium cations **5.23** or **5.24**, the most stable radical R$^{1\cdot}$ or R$^{2\cdot}$ being lost preferentially. These acylium cations (**5.25**) are subject to mesomeric stabilization.

$$RC\equiv O^+ \leftrightarrow R\overset{+}{C}=O$$

(5.25)

For aldehydes, the loss of H· is less favoured than that of R·, so the fragment **5.26** (m/z 29) is commonly observed in their mass spectra. Ketones give fragments in the series $C_nH_{2n+1}CO^+$ (m/z 43, 57, 71, etc.), which unfortunately have the same nominal m/z values as the $C_nH_{2n+1}^+$ ions; high-resolution spectrometry and accurate mass determination are required to separate and distinguish between these isobaric ions (e.g. $C_2H_5CO^+$ and $C_4H_9^+$, both m/z 57). These acylium cations (**5.25**) undergo further simple fission with the loss of carbon monoxide, thus entering the isobaric series of $C_nH_{2n+1}^+$ ions.

$$R-\overset{\overset{O^{+\cdot}}{\|}}{C}-H \longrightarrow R^{\cdot} + HC\equiv O^+$$
$$m/z = 29$$
(**5.26**)

$$R-C\equiv O^+ \rightarrow R^+ + CO$$
(**5.25**)

The presence of a γ-hydrogen atom in either R^1 or R^2 in structure **5.22** leads to the McLafferty rearrangement, in which an alkene is lost during the formation of **5.27** via a hydrogen transfer. The process is believed to involve a cyclic six-membered transition state. The resultant odd-electron ions **5.26** are normally abundant; the lowest m/z value, 44, corresponds to the rearrangement of an aldehyde ($R^2 = H$) with no α-substituents. When the alkene fragment is able to support a positive charge, then this rearrangement may result in a $C_nH_{2n}^{+\cdot}$ ion (**5.28**). For example, substitution by an aromatic unit on either the β- or γ-carbon will greatly enhance this tendency.

m/z 44 + (n×14)

(**5.27**)

m/z M — [44+(n×14)]

(**5.28**)

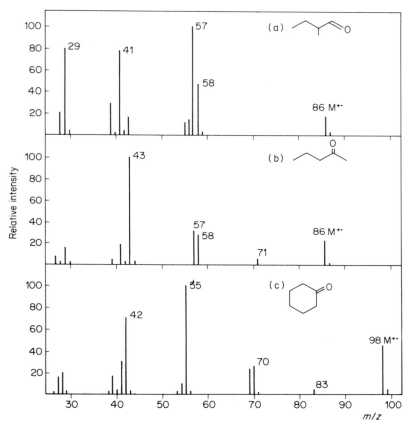

Figure 5.13 70 eV EI spectra of (a) 2-methylbutanal, (b) pentan-2-one and (c) cyclohexanone.

The spectra in Fig. 5.13 illustrate some aspects of these fragmentations. The spectrum of 2-methylbutanal (Fig. 5.13a) shows a strong m/z 29 peak, although high-resolution studies show that both isobaric ions $C_2H_5^+$ and HCO^+ are involved. The base peak at m/z 57 corresponds to the secondary carbonium ion $C_4H_9^+$. The McLafferty rearrangement ion at m/z 58, which involves the transfer of a primary hydrogen, is less prominent. The spectrum of pentan-2-one (Fig. 5.13b) shows ions at m/z 43 and 71 corresponding to the two acylium cations arising from cleavage adjacent to the carbonyl function. Again, high-resolution examination reveals that the m/z 43 peaks is a doublet consisting of both CH_3CO^+ and $C_3H_7^+$ ions. The McLafferty rearrangement ion at m/z 58 is significant, although again not especially abundant in this spectrum.

5.5.1.8 Cyclic ketones

Again, the cyclic structure requires a minimum of two bond cleavages to change the m/z value; hence the molecular ion tends to be more abundant than for similar acyclic ketones. Typical fragmentations are illustrated by the spectrum of cyclohexanone (Fig. 5.13c).

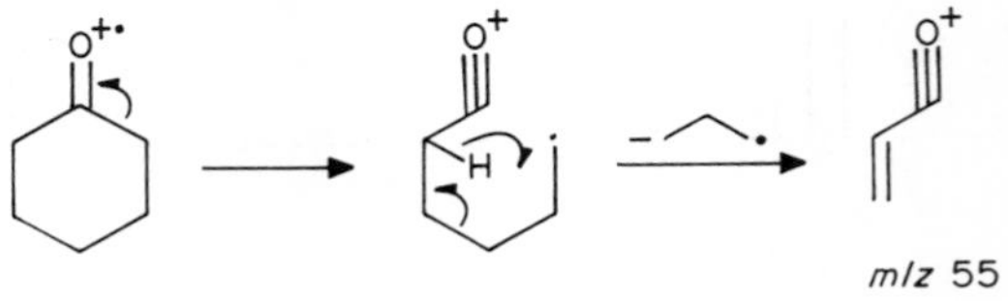

m/z 55

(5.29)

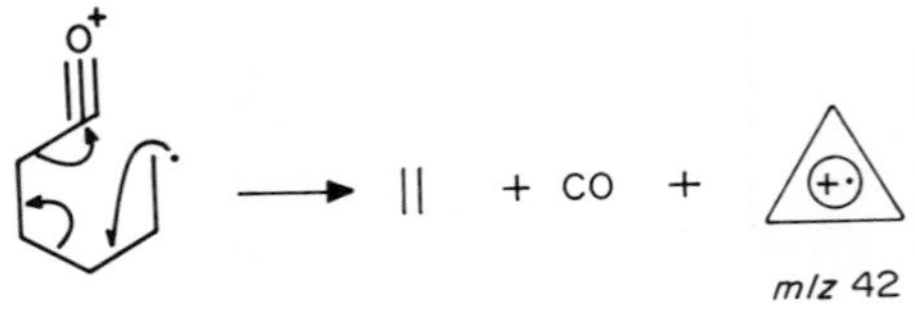

(5.30)   (5.31)

m/z 42

The ring cleavage, adjacent to the carbonyl group, is typically followed by hydrogen migration and loss of an alkane radical to yield an ion (**5.29**) at m/z 55. The intermediate ion **5.30** may undergo loss of carbon monoxide alone to yield an alkene equivalent at m/z 70 or concurrent with ethene to yield another alkene equivalent at m/z 42, which is often written as the cyclic ion-radical **5.31**. The low abundance ion at m/z 83, corresponding to expulsion of a methyl radical from the molecular ion, indicates that complicated rearrangements and fragmentation can occur during a high-energy process, such as electron bombardment.

5.5.1.9 Acetals and ketals

These common derivatives of aldehydes and ketones give rise to mass spectra, in which the molecular ion is generally very weak or even absent. Strong peaks from cleavage of one or other of the bonds to the central carbon atom, which result in ions (e.g. **5.33**, **5.34** or **5.35**) stabilized by oxonium ion formation.

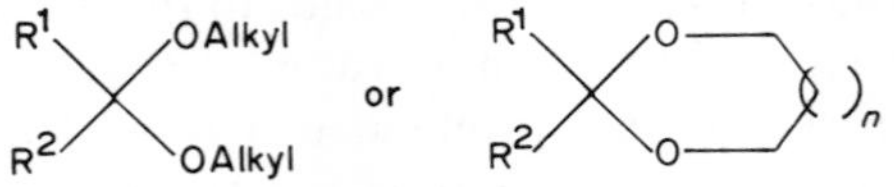

(5.32)

Acetals: $R^1 =$ H, $R^2 =$ alkyl or aryl, $n = 0$ or 1
Ketals: $R^1 = R^2 =$ alkyl or aryl, $n = 0$ or 1

(5.33) (5.34) (5.35)

5.5.1.10 Carboxylic acids, esters and amides

As expected, the presence of a carbonyl oxygen provides a convenient site for a charged-localized molecular ion (5.36). This species again permits alternative simple fissions, adjacent to the carbonyl group, yielding either the ion 5.37 or the acylium cation 5.25, respectively. The nature of the group X tends to control which process is favoured. Thus when X = OH or NH_2 process a is often observed, leading to the ions $HOC\equiv O^+$ (m/z 45) or $NH_2C\equiv O^+$ (m/z 44), respectively, whereas X = OR, NHR or NR_2 more commonly yield acylium cations (5.25). As usual, these tend to lose carbon monoxide giving rise to alkyl carbonium ions.

$$R^\bullet + \overset{+}{O}\equiv CX$$
(5.37)

$$RC\equiv O^+ + X^\bullet$$
(5.25)

$$R^+ + CO$$

(5.36)

The presence of a γ-hydrogen atom relative to the carbonyl oxygen, as before, leads to McLafferty rearrangement, when loss of an alkene results in ions (5.38) belonging to a homologous series $C_nH_{2n}O_2^{+\bullet}$ ($n \geqslant 2$) for acids and esters, and $C_nH_{2n+1}NO^{+\bullet}$ ($n \geqslant 2$) for amides. In the case of esters and secondary or tertiary amides a second pathway for McLafferty rearrangement may exist within the group X. Thus, as before, alkenes can be lost leading to the rearrangement ion 5.39 in the same homologous series $C_nH_{2n}O_2^{+\bullet}$ or $C_nH_{2n+1}NO^{+\bullet}$, although in this case n = 1 is possible for formates. As with the similar processes of aldehydes and ketones, the charge may be retained in the alkene fragment, if it is well stabilized.

X = OH m/z 60 + (n × 14)
X = OR m/z 74 + (n × 14)
X = NR_2 m/z 59 + (n × 14)

(5.38)

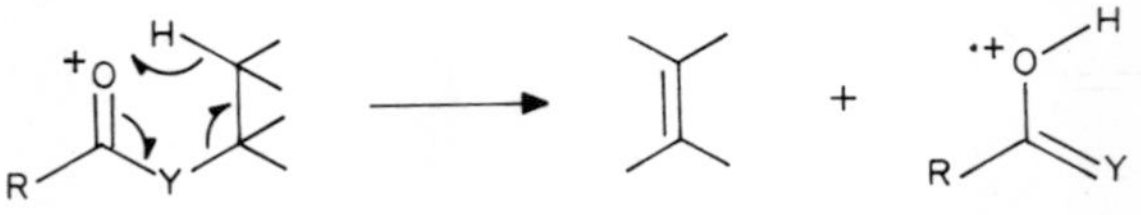

$$Y = O, \quad m/z\ 46 + (n \times 14)$$
$$Y = NR, \quad m/z\ 45 + (n \times 14)$$

(5.39)

Ethyl or higher alkyl esters sometimes show a peak due to an ion **(5.40)** formed by a double rearrangement process. These even-electron ions have m/z values which can be characteristic of the acid (e.g. formates $m/z = 47$, acetates $m/z = 61$, propanoates $m/z = 75$).

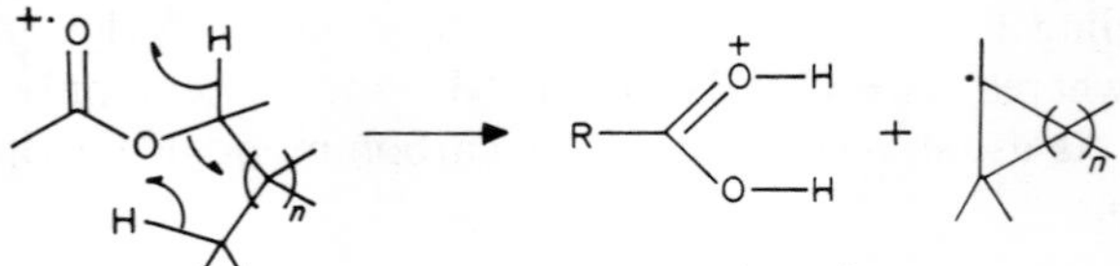

$$m/z\ 47 + (n \times 14)$$

(5.40)

Some of these fragmentations are illustrated by the mass spectra in Fig. 5.14. The mass spectrum of butyl acetate (Fig. 5.14a) shows a McLafferty type of rearrangement, where the charge is retained on the alkene fragment **(5.41)**. There

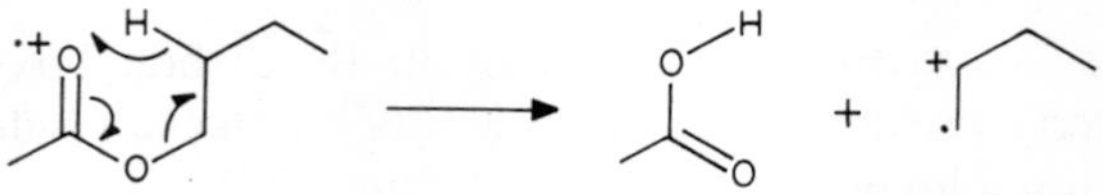

$$m/z\ 56$$

(5.41)

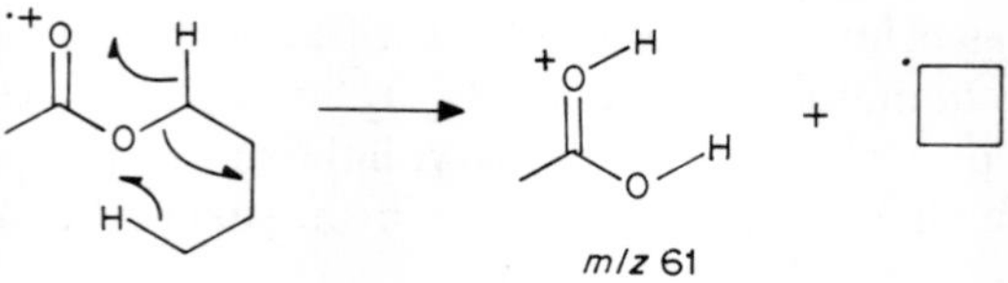

$$m/z\ 61$$

(5.42)

is also a significant double rearrangement ion **(5.42)** at m/z 61, indicating an acetate ester. The remaining ions are of an alkyl or alkenyl nature. The spectrum of 2-ethylbutanoic acid (Fig. 5.14b) shows a strong McLafferty rearrangement ion **(5.43)** at m/z 88, which pinpoints the 2-ethyl substitution. The strong even-electron fragment **5.44** at m/z 73 may reflect the ready loss of a methyl radical from **5.43**, while ions at m/z 45 (**5.37**, X = OH) and m/z 71 ($M - 45$; $C_5H_{11}^+$) confirm the presence of an acid function. It is surprising that m/z 43 ($C_3H_7^+$) should constitute the base peak, since its formation requires at least hydrogen

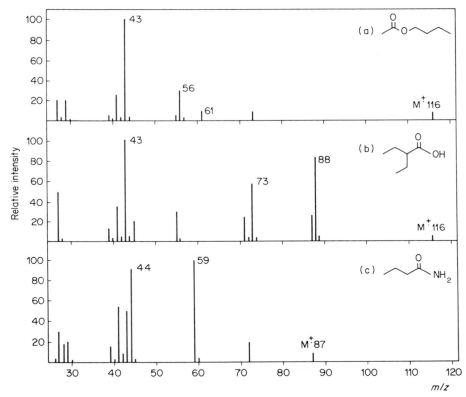

Figure 5.14 70 eV EI spectra of (a) butyl acetate, (b) 2-ethylbutanoic acid and (c) butanamide.

rearrangement. Indeed, this spectrum illustrates why not too much importance should be placed upon small alkyl fragments when deciding the structure of an unknown. The molecular ion for this branched chain acid is very weak. Butanamide (Fig. 5.14c) is marked as a primary amide by the presence of the ion (**5.37**, X = NH$_2$) at m/z 44 in its mass spectrum. Finally, the existence of an abundant McLafferty rearrangement ion (**5.45**) at m/z 59 completely determines the structure.

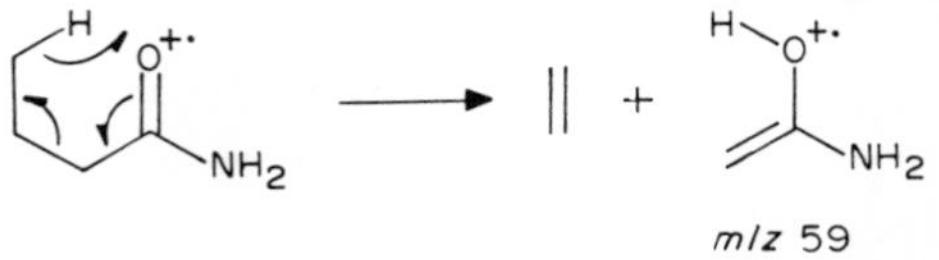

[diagram: McLafferty rearrangement giving m/z 59]

(5.45)

5.5.1.11 Haloalkanes

The very obvious isotopic patterns are characteristic of the chloro- and bromoalkanes (see Section 5.3.4). The lone pairs of electrons on the halogen atoms again provide a low-energy site for a charge-localized molecular ion (**5.46**).

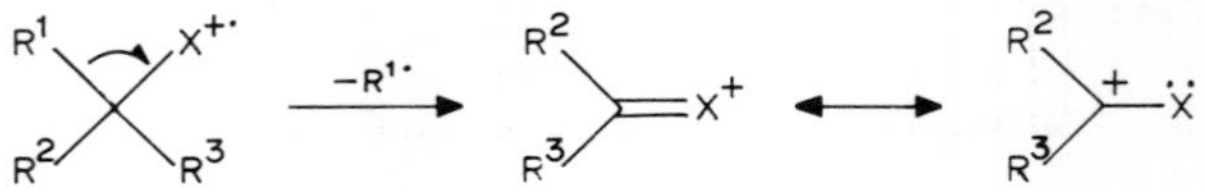

[diagram showing (5.46) → loss of $R^{1\cdot}$ → resonance structures]

(5.46) $X = F, m/z\ 33 + (n \times 14)$

(5.47)

Simple fission adjacent to the halogen could result in a resonance-stabilized ion (**5.47**). This type of cleavage is generally of little importance, however, with chlorine, bromine or iodine, possibly owing to the poor π-bonds formed by second-row and higher elements on overlap with 2p orbitals, which means only poor mesomeric stabilization in **5.47**. Compared with alcohols, ethers, etc., even fluorine, which has lone pairs of electrons available in the appropriate 2p orbitals, does not lead to abundant simple fission ions of the type **5.47** ($X = F, m/z\ 33$). In this case, the high electronegativity of fluorine probably reduces the effective stabilization due to resonance. The C—X bond in haloalkanes may be cleaved in two ways. Heterolysis of the bond (process a) can be seen in the mass spectra of all halides, but homolysis (process b) is only observed with the less electronegative iodides and to a lesser extent bromides. For the more electronegative fluorine and chlorine compounds, elimination of hydrogen halide is often more important, leading to odd-electron ions at $m/z\ M - 20$ and $M - 36/38$, respectively. The loss of the chlorine atom is noticeable owing to the change in isotopic ratios associated with the $M - 36/38$ peak compared with the molecular-ion cluster.

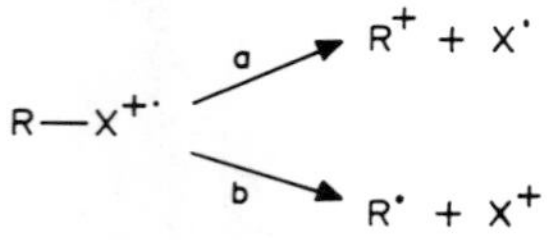

[diagram: $R-X^{+\cdot}$ → (a) $R^+ + X^\cdot$; (b) $R^\cdot + X^+$]

It is possible to summarize the main effects of the various halogens in haloalkane spectra as outlined below.

Alkyl fluorides

The loss of HF to yield a peak $M - 20$ is the main fragmentation involving fluorine.

Alkyl chlorides

The elimination of HCl is still the major process leading to a peak $M - 36/38$, but it is usually possible to detect a less abundant peak $M - 35/37$ due to loss of Cl˙.

Alkyl bromides

The fragmentations are similar to the chloro compounds, although now loss of Br˙ ($M - 79/81$) is usually preferred to elimination of HBr ($M - 80/82$). Weak Br^+ peaks at m/z 79/81 can sometimes be detected.

Alkyl iodides

Loss of I˙ is the most common fragmentation ($M - 127$), and there is usually a noticeable I^+ peak at m/z 127.

The mass spectra in Fig. 5.15 (see p. 170) illustrate this summary.

5.5.2 Aromatic compounds

The stabilizing effect of the aromatic π-electron system and the difficulty of cleaving a benzenoid ring mean that the molecular ion is usually observed, and often abundant, in the spectra of this class of compound. This stability may also lead to the removal of a second electron in the ion source, resulting in doubly charged ions. Since these are focused according to their m/z ratio, then for uneven values of m peaks at half mass units are observed. Such doubly charged ions are an indication of aromatic (or heteroaromatic) compounds.

Once the benzenoid ring is cleaved, low-abundance ions result in the series $C_nH_n^{+\cdot}$, which may arise due to successive loss of acetylene. In addition, loss of one or two hydrogens from these ions may be observed. Thus aromatic compounds are characterized by weak peaks in the groups m/z 78, 77, 76; 65, 64, 63; 52, 51, 50; and 39, 38, 37.

5.5.2.1 Simple fission adjacent to an aromatic ring

This is a common fragmentation, although not often leading to abundant ions, as the resultant phenonium or substituted phenonium ions (**5.48**) are not generally especially stable; X may be NO_2, OR or halogen.

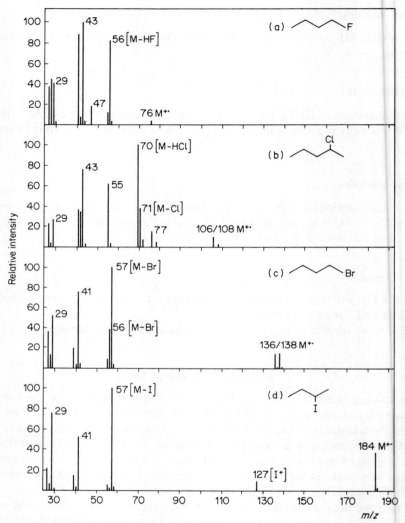

Figure 5.15 70 eV EI spectra of (a) 1-fluorobutane, (b) 2-chloropentane, (c) 1-bromobutane and (d) 2-iodobutane.

(5.48)

5.5.2.2 Simple fission of a bond αβ to an aromatic ring

This is often the most important fragmentation in aromatic compounds. This type of cleavage generates a positive charge adjacent to the aromatic π-electron system, which leads to resonance-stabilized ions. In consequence, these are generally much more abundant than the less stable phenonium ions.

(5.49) → (5.50) X=H, m/z 91 → (5.51) X=H, m/z 65

The mass spectra of alkylbenzenes (5.49) exhibit very abundant ions due to this type of cleavage, which can best be represented (Section 5.4.4) as a tropylium ion (5.50). Further loss of acetylene ensues, giving the ion 5.51 (m/z 65). Aromatic ethers similarly yield ions of the type 5.52, which readily eliminate carbon monoxide, again with the formation of the ion 5.51 (m/z 65). Nitro compounds, in addition to losing $NO_2^{\cdot}$ ($M-46$), leading to ions of the type 5.48, also rearrange under electron bombardment to give nitrites (5.53), which then lose $NO^{\cdot}$ ($M-30$) by αβ-bond cleavage, resulting in oxonium ions of the type 5.52. Aromatic aldehydes, ketones, acids and esters all undergo a form of αβ-bond cleavage with loss of an X radical, the resultant benzoyl cations 5.54 usually being abundant in their mass spectra. Loss of carbon monoxide follows to yield phenonium ions (5.48).

(5.52) (5.53) (5.51) X=H, m/z 65

(5.54) Y=H, m/z 105 (5.48) Y=H, m/z 77

5.5.2.3 McLafferty rearrangement

Where a γ-hydrogen atom exists with respect to the ring, abundant fragment ions often result by rearrangement via a six-membered cyclic transition state of the aromatic ions. Examples are shown below.

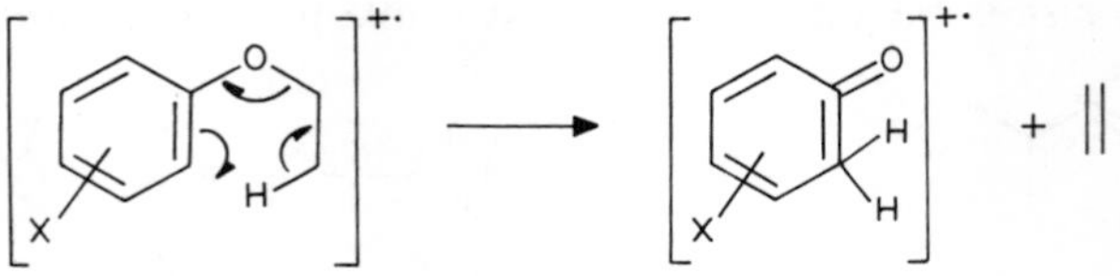

X=H, m/z 94

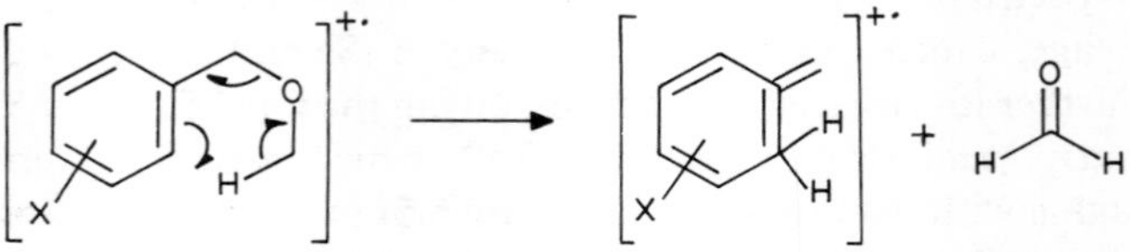

X=H, m/z 92

5.5.2.4 Other rearrangements

A number of aromatic molecular ions undergo rearrangements involving four-membered cyclic transition states; the most common example of this type is the loss of formaldehyde $(M - 30)$ from methoxy derivatives (**5.55**). Evidence that this type of process arises from the molecular ion of anisole (**5.55**, X = H) in a one-step reaction is provided by the presence of a metastable peak in its mass spectrum at m/z 56.3 $[m^* = (78)^2/108 = 56.3]$. Aromatic sulphonamides (**5.56**) readily lose sulphur dioxide, involving a rearrangement which can also be represented as occurring *via* a four-membered transition state. The resultant anilinium ion **5.57** undergoes further rearrangement with loss of HCN. Indeed, the loss of hydrogen cyanide from anilines is another common process in aromatic systems. In many respects this elimination is similar to the expulsion of carbon monoxide from phenols (see below), in that it requires a preliminary rearrangement to the imino tautomer (**5.58**) before loss of HCN.

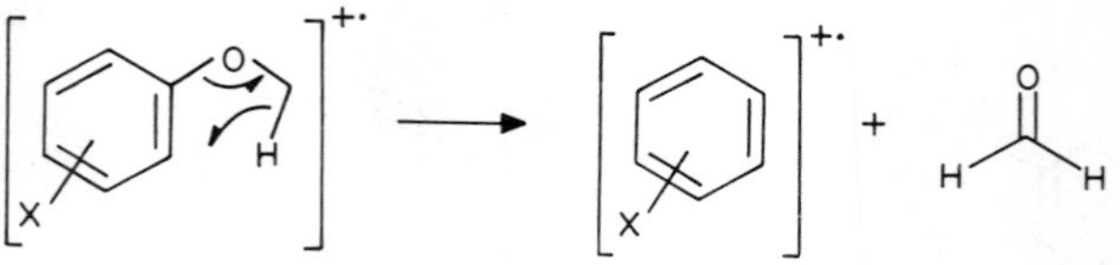

(**5.55**)   X=H, m/z 78

[Structures 5.56 → 5.57 + SO$_2$, with X=H, m/z 93]

[Structure 5.57 → 5.58 (X=H, m/z 93) → −HCN → cyclopentadienyl cation (X=H, m/z 66)]

Rearrangement with the loss of carbon monoxide is also common in aromatic systems. This has already been noted for the ion **5.52** derived from aromatic ethers of nitro compounds. In this type of rearrangement hydrogen is not transferred, instead there is a carbon skeletal shift. The spectrum of fluorenone provides another example: the very strong peak at m/z 152, which arises by expulsion of carbon monoxide and the formation of a biphenylene ion (**5.59**).

[Fluorenone m/z 180 → −CO → (5.59) m/z 152]

The carbonyl group does not always need to be present in the substrate, since rearrangement under electron bombardment may lead to its formation. Thus, phenols may eliminate carbon monoxide, after rearrangement to the keto-isomer (**5.60**).

[Phenol → keto isomer (5.60) X=H, m/z 94 → −CO → cyclopentadienyl cation X=H, m/z 66]

5.5.2.5 Elimination due to ortho-groups

The presence of *ortho*-substituents may lead to the ready elimination of a neutral molecule e.g. H_2O, ROH, RCO_2H, etc. Thus, elimination of alcohol is observed with the *ortho*-substituted esters (**5.61**); X = O, NH or CH$_2$, although it is least favourable when X = CH$_2$.

(**5.61**)

5.5.2.6 Retro Diels–Alder reactions

Whenever a compound contains a double bond in a six-membered ring, which might have been formed in a Diels–Alder reaction, then the high energy of its excited molecular ion favours an intra-molecular dissociative process, i.e. a Retro Diels–Alder reaction. The double bond involved does not require to be isolated, but can even form part of an aromatic (or hetero-aromatic) system. For example, the molecular ion (**5.62**) derived from tetralin, loses ethene with concomittant formation of the radical cation (**5.63**) in such a process.

(**5.62**)   (**5.63**)

5.5.3 Heteroaromatic compounds

Heteroaromatic compounds resemble their carbocyclic analogues in that they give rise to stable molecular ions and the possibility of doubly charged ions. Cleavage of the aromatic ring in benzene derivatives tends to proceed with ejection of acetylene, whereas heteroaromatics cleave with preferential loss of the heteroatom in a neutral species. Thus pyridines and pyrroles tend to lose HCN, and thiophenes and furans lose CHS˙ and CHO˙, respectively. These processes, as with aromatic compounds, result in ions of the general composition $C_nH_n^{+\cdot}$; as before (see Section 5.2.2), further loss of one or two hydrogens from these ions may be observed. Thus heteroaromatic compounds often show the same weak peaks in the groups m/z 65, 64, 63; 52, 51, 50; and 39, 38, 37 as do benzene derivatives.

It has been shown, by deuterium labelling experiments, that benzene derivatives often undergo extensive rearrangement before fragmentation. Similar rearrangement can occur in heteroaromatic compounds. Thus, the mass spectra of oxazoles (**5.64**) and isoxazoles (**5.65**) show many features in common, suggesting that they may come into equilibrium under electron bombardment. This process could involve an azirine intermediate (**5.66**).

In other respects, heteroaromatic compounds exhibit similar side-chain and substituent fragmentations to those described above for aromatic compounds (see Section 5.5.3).

$$[\text{(5.64)}] \rightleftharpoons [\text{(5.66)}] \rightleftharpoons [\text{(5.65)}]$$

5.6 INTERPRETATION OF A MASS SPECTRUM

It is difficult to define a precise procedure for the interpretation of a mass spectrum; in fact, every example requires almost individual treatment. This is not surprising, when one considers the large range of organic structures and the wide variety of gas-phase unimolecular processes by which they can fragment. Indeed, the above discussion (see Section 5.5) of fragmentations associated with functional groups is necessarily brief, for reasons of space. The following procedure is suggested as a guide to the beginner.

i. Note the conditions under which the spectrum was obtained. Predict the effect on the spectra, where these vary from normal.

ii. Note any information available from alternative sources. It is unusual to determine a structure using mass spectral evidence alone. Thus, physical constants, method of preparation, chemical reactions of the unknown and other spectral data may all greatly assist mass spectral interpretation.

iii. Find the molecular-ion cluster (Section 5.3.1). Where the molecular ion may be weak, confirmation should be sought using 'soft ionization' techniques (CI, FAB, etc.; see Section 5.3.2) to reveal a pseudo-molecular ion $(M + 1)^+$. Care should also be exercised if the presence of impurities is suspected; two commonly encountered impurities are phthalate plasticizers ($m/z = $ 149, 167, 279) and silicone grease ($m/z = $ 133, 207, 281, 355, 429). It is perhaps worth noting that the mass spectrum observed depends on the partial pressure of a component of a mixture in the source. Thus even a minor component can, if it is more volatile, dominate the spectrum.

iv. Determine the molecular formula. This can be obtained from the molecular ion by application of accurate high-resolution mass measurement (Section 5.3.3), consideration of natural isotropic abundances (Section 5.3.4), use of elemental analysis figures, proton counts from an NMR spectrum, etc. Two simple rules may help on occasions:

(a) A molecular ion with an even m/z value contains either no nitrogen atoms or an even number of nitrogen atoms. Conversely, a molecular ion with an odd m/z value contains an odd number of nitrogen atoms.

Table 5.3. Common neutral fragments Lost from molecular Ions $(M - X)$

Mass lost (X)	Composition	Possible inference (R = aryl or alkyl)
1	$H^{\cdot}$	—
2	H_2	—
14	$CH_2^{+\cdot}$	Homologous impurity
15	$CH_3^{\cdot}$	—
16	$NH_2^{\cdot}$	$RCONH_2, RSO_2NH_2$
	O	$RNO_2, R_3NO, R_2SO, R_2SO_2$
17	$OH^{\cdot}$	$RCO_2H, RNO_2, R_3NO, RNHOH, R_3COH$
18	H_2O	Aliphatic alcohols, aldehydes, ketones and carboxylic acids
19	$F^{\cdot}$	RF
20	HF	Alkyl fluorides
26	C_2H_2	Aromatic compounds
	$CN^{\cdot}$	RCN
27	HCN	N-heteroaromatics
28	C_2H_4	McLafferty rearrangements and retro-Diels–Alder reactions
	CO	Quinones, fluorenones, HCO_2R
29	$C_2H_5^{\cdot}$	—
	$CHO^{\cdot}$	ArOH, RCHO, O-heteroaromatics
30	CH_2O	$AROCH_3$
	$NO^{\cdot}$	RNO_2
31	CH_3NH_2	Alkylmethylamines
	$CH_3O^{\cdot}$	RCO_2CH_3
32	CH_3OH	Methyl alkanoates
	S	R_2S
33	$HS^{\cdot}$	RSH
34	H_2S	Alkanethiols
35/37	$Cl^{\cdot}$	RCl
36/38	HCl	Alkyl chlorides
41	$C_3H_5^{\cdot}$	Alkenes, alicyclics
42	C_3H_6	$RCOC_4H_9, ArC_4H_9$
	CH_2CO	CH_3CO_2R, CH_3CONHR
43	$C_3H_7^{\cdot}$	$ArC_4H_9, RCOC_3H_7$, alkanes
	$CH_3CO^{\cdot}$	$RCOCH_3$
44	CO_2	$RCO_2H, RCO_2R, (RCO)_2O$
45	$C_2H_5O^{\cdot}$	$RCO_2C_2H_5$
	$CO_2H^{\cdot}$	RCO_2H
46	C_2H_5OH	Ethyl alkanoates
	$NO_2^{\cdot}$	RNO_2
47	$CH_3S^{\cdot}$	$RSCH_3$
48	CH_3SH	Alkyl methyl sulphides
	SO	R_2SO, R_2SO_2
55	$C_4H_7^{\cdot}$	Alkenes
56	C_4H_8	$ArC_5H_{11}, ArOC_4H_9, RCOC_5H_{11}$ and tetralins (retro-Diels–Alder)
57	$C_4H_9^{\cdot}$	Alkanes, $RCOC_4H_9, ArC_5H_{11}$
	$C_2H_5CO^{\cdot}$	Alkyl ethyl ketones
60	CH_3CO_2H	Alkyl ethanoates

Table 5.4. Some Common fragment ion structures

m/z	Composition	Possible inference (R = aryl or alkyl)
19	F^+	RF
20	$HF^{+\cdot}$	Alkyl fluorides
27	$C_2H_3^+$	—
28	$C_2H_4^{+\cdot}$	—
	$CO^{+\cdot}$	—
	$N_2^{+\cdot}$	—
29	$C_2H_5^+$	—
	CHO^+	RCHO
30	$CH_2{=}NH_2^+$	Primary alkylamines
31	$CH_2{=}OH^+$	Primary alkyl alcohols
36/38	$HCl^{+\cdot}$	Alkyl chlorides
39	$C_3H_3^+$	Aromatics
41	$C_3H_5^+$	Alkenes, cycloalkenes
43	$C_3H_7^+$	Alkanes, C_3H_7X
	CH_3CO^+	CH_3COX
44	$CH_3CH{=}NH_2^+$	Alkyl-$CH(CH_3)NH_2$
	$CH_2{=}NHCH_3^+$	Alkyl-$NHCH_3$
	$CH_2{=}CHOH^{+\cdot}$	Alkyl-CHO (McLafferty)
	$NH_2C{\equiv}O^+$	$RCONH_2$
	$CO_2^{+\cdot}$	RCO_2H
45	$CH_3CH{=}OH^+$	Alkyl-$CH(CH_3)OH$
	$CH_2{=}OCH_3^+$	Alkyl-CH_2OCH_3
	CO_2H^+	RCO_2H
47	$CH_2{=}SH^+$	Primary alkanethiols
49/51	$CH_2{=}Cl^+$	Primary alkyl chlorides
50	$C_4H_2^{+\cdot}$	} Aromatics
51	$C_4H_3^+$	} Heteroaromatics
55	$C_4H_7^+$	Alkenes
57	$C_4H_9^+$	Alkanes
58	$C_3H_8N^+$	Some alkylamines
	$CH_2{=}C(CH_3)OH^{+\cdot}$	Alkyl-$COCH_3$ (McLafferty)
59	$C_2H_5CH{=}OH^+$	Alkyl-$CH(C_2H_5)OH$
	$CH_2{=}OC_2H_5^+$	Alkyl-$CH_2OC_2H_5$
	$CH_2{=}C(NH_2)OH^{+\cdot}$	Alkyl-$CONH_2$ (McLafferty)
	$CO_2CH_3^+$	RCO_2CH_3
60	$CH_2{=}C(OH)_2^{+\cdot}$	Alkyl-CO_2H (McLafferty)
61	$CH_3C(OH){=}OH^+$	$CH_3CO_2C_nH_{2n+1}$ ($n \geqslant 2$; double rearrangement)
	$\overset{+}{S}H$ $\underset{CH_2{-}CH_2}{\diagup\diagdown}$	XCH_2CH_2SH
64	$C_5H_4^{+\cdot}$	} Aromatics, heteroaromatics
65	$C_5H_5^+$	
66	$H_2S_2^{+\cdot}$	RSSR
69	$C_5H_9^+$	Alkenes
71	$C_5H_{11}^+$	Alkanes
72	$C_4H_{10}N^+$	Some alkylamines
	$C_4H_8O^{+\cdot}$	Aldehydes/ketones (McLafferty)
73	$CO_2C_2H_5^+$	$RCO_2C_2H_5$
74	$C_3H_6O_2^{+\cdot}$	Carboxylic acids/methyl esters (McLafferty)
75	$C_2H_5C(OH){=}OH^+$	$C_2H_5CO_2C_nH_{2n+1}$ ($n \geqslant 2$; double rearrangement)
76	$C_6H_4^{+\cdot}$	}
77	$C_6H_5^+$	} Aromatics
78	$C_6H_6^{+\cdot}$	}

Table 5.4 (*Contd.*)

m/z	Composition	Possible inference (R = aryl or alkyl)
79/81	Br^+	RBr
80/82	$HBr^{+\cdot}$	Alkyl bromides
80	$C_5H_6N^+$	Alkylpyrroles
81	$C_5H_5O^+$	Alkylfurans
91	$C_7H_7^+$	$C_6H_5CH_2X$
92	$C_6H_6N^+$	Pyridyl-CH_2X
105	$C_6H_5CO^+$	Phenyl ketones, benzoate esters
122	$C_6H_5CO_2H$	Alkyl benzoates

    (b) A molecule containing nitrogen, phosphorus or halogen atoms has an even number of hydrogen atoms accompanying an even-numbered combination of these atoms. An odd-numbered combination of these elements is accompanied by an odd number of hydrogen atoms.

v. Determine the double-bond and ring equivalent from the molecular formula using the equation

$$\tfrac{1}{2}(2n_4 + n_3 - n_1 + 2) = \text{number of double bonds and/or rings present}$$

where n_4 = number of tervalent atoms (C, Si), n_3 = number of trivalent atoms (N, P), n_1 = number of univalent atoms (H, halogens) in the molecular formula. (Note that a triple bond, as in acetylene or a nitrile, counts as two double bond equivalents and a benzenoid ring counts as four.)

vi. Note the intensity of the molecular ion and the general disposition of ions within the spectrum. This may allow the class of compound to be inferred (Section 5.3.1).

vii. List all important fragments ions in the spectrum, noting even- and odd-electron ions (Section 5.4.3). Look for even-electron ion series (e.g. $C_nH_{2n}X^+$), which may reveal the nature of the functional group (X). Odd-electron ions often have structural significance, since they correspond in general to rearrangement or elimination ions.

viii. Note the difference in m/z values between the molecular ion and higher mass fragments, and between the fragments themselves. Identify possible neutral fragments lost and confirm with metastable peaks or linked scans (Sections 5.4.1 and 5.4.2), where these were observed. Table 5.3 may be helpful in this connection.

ix. Postulate structures for significant fragments ions; use could be made of Table 5.4. Confirm using any accurate high-resolution mass data or if possible natural isotopic abundances. The latter require the absence of other fragment ions, which have m/z values equal to those of the isotopic fragment peaks, and is at best a rough guide to fragment ion composition. The presence of atoms with abundant natural isotopes (e.g. chlorine, bromine, sulphur) can usually be readily detected.

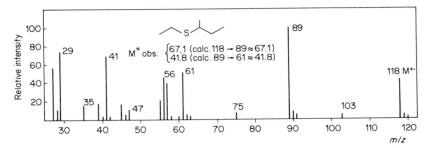

Figure 5.16 70 eV EI spectrum for unknown 1.

x. Assemble the fragment structures and deduce possible molecular structures. Test the conclusion against all known data. Final confirmation requires comparison of the mass spectrum with that of an authentic sample. However, it should be remembered that in contrast to the spectroscopic methods (UV, IR, NMR), mass spectra run on different instruments may show considerable variation in the intensities of the peaks observed. In fact, it is not unknown for different peaks to be observed due to, say, thermal decomposition in the inlet system of only one of the two instruments. Therefore, it may be advisable to confirm the structure by comparison of the more reproducible IR or NMR spectra.

5.7 EXAMPLES OF THE INTERPRETATION OF MASS SPECTRA

5.7.1 Unknown 1

The spectrum in Fig. 5.16 was obtained at 70 eV under EI conditions. The molecular ion peak at m/z 118 is accompanied by isotope peaks at m/z 119 and 120. The intensity ratios of M, $M + 1$ and $M + 2$ peaks are 100:7.51:4.66; the relatively high $M + 2$ value suggests the presence of a sulphur atom, while the $M + 1$ value requires around six carbon atoms. The even numbered value for M means that either no or an even number of nitrogen atoms are present. This suggests a molecular formula of $C_6H_{14}S$ (Section 5.3.4), confirmed by the observed accurate mass 118.0809 (calculated, 118.0813), the error of -3.3 ppm being within the accepted limit of ± 10 ppm (Section 5.3.3).

The double-bond and ring equivalent $= \frac{1}{2}[(2 \times 6) - 14 + 2] = 0$ [see Section 5.6(v)]; we can therefore conclude that the unknown is an acyclic thiol or thioether. Examination of the spectrum (Fig. 5.16) reveals that there are no peaks at m/z 84 $(M - H_2S)^{+\cdot}$ or 85 $(M - HS)^+$, so the unknown is a thioether. Three possibilities ($CH_3SC_5H_{11}$, $C_2H_5SC_4H_9$ or $C_3H_7SC_3H_7$) exist for the distribution of carbon atoms about the sulphur atom. The fact that m/z 89 constitutes the base peak strongly suggests loss of an ethyl radical by simple fission of a bond β to the sulphur atom. Only three structures (**5.67, 5.68** and **5.69**) satisfy this requirement. A fourth structure (**5.70**), although containing a suitable ethyl group, would more

(**5.67**)   (**5.68**)   (**5.69**)

(**5.70**) → *m/z* 103 (**5.71**)

readily lose a methyl radical at the alternative secondary site, yielding the ion **5.71** at *m/z* 103, which is a relatively weak peak in the spectrum. The molecular ion yields the fragment ion of *m/z* 89, which can readily eliminate ethene to form an ion of *m/z* 61; the metastable ions ($m^* = 67.1$ and 41.8) confirm these processes. This latter process is possible for either structure **5.67** or **5.68**, but is unlikely for **5.69**, which may be expected to eliminate propene yielding a stronger peak at *m/z* 47.

(**5.67**) → *m/z* 89 → $-C_2H_4$ → *m/z* 61

(**5.68**) → *m/z* 89 → $-C_2H_4$ → *m/z* 61

(**5.69**) → *m/z* 89 → $-C_3H_6$ → *m/z* 47

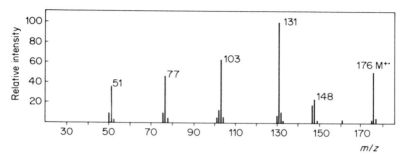

Figure 5.17 70 eV EI spectrum for unknown 2.

Finally, the presence of strong peaks at m/z 57, 56 and 29 ($C_4H_9^+$, $C_4H_8^+$ and $C_2H_5^+$) fits well with a C_2H_5—S—C_4H_9 structure rather than CH_3—S—C_5H_{11}, in which cleavage of a C—S bond can be expected to give m/z 71 ($C_5H_{11}^+$); this latter peak is not observed. In a practical situation, the sample would now be compared with authentic *sec*-butyl ethyl sulphide (**5.68**).

5.7.2 Unknown 2

The relatively strong molecular ion (Fig. 5.17) under normal 70 eV EI conditions, plus important peaks at m/z 51 and 77 (possibly $C_4H_3^+$ and $C_6H_5^+$) suggest an aromatic structure. The molecular formula can be assigned as $C_{11}H_{12}O_2$ (calculated, 176.0834) by accurate mass measurement. The double bond equivalent and ring equivalent is $1/2[(2 \times 11) - 12 + 2] = 6$, which again strongly supports an aromatic structure, which is necessary to account for the relatively high value.

The base peak (m/z 131) corresponds to a loss of 45 mass units, possibly $C_2H_5O^{\cdot}$, which taken in connection with the odd-electron peak at m/z 148 ($M - 28$) suggests and ethyl ester and McLafferty rearrangement (**5.72**). The further loss of carbon monoxide from the base peak to yield an abundant ion of m/z 103 supports this conclusion. The probable presence of an unsubstituted phenyl group is suggested by the peaks at m/z 77 and 51, which leads to a part structure **5.73**. However, it is not easy to distinguish between the 1-carbethoxy-1-phenylethene **5.75** and the E or Z isomers of the cinnamic esters **5.74** by a mass spectrum alone. Indeed, even comparison with the spectra from authentic samples may not provide a convincing proof of structure for the (E) and (Z)-cinnamates. The final proof that this sample was the E isomer of ethyl cinnamate would be best provided by an NMR spectrum or comparison of the IR spectrum with that of an authentic sample.

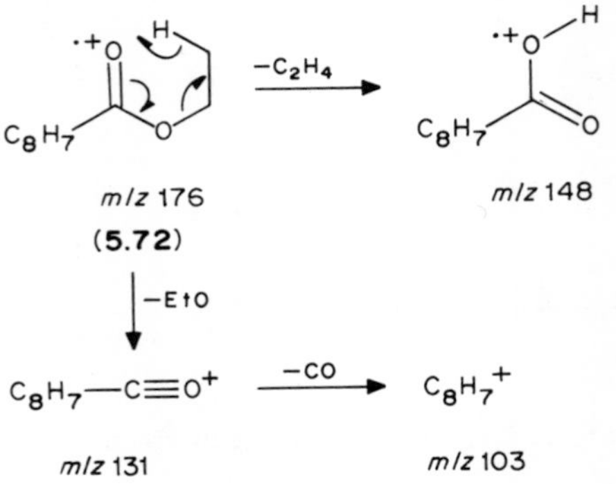

m/z 176
(5.72)

m/z 148

m/z 131

m/z 103

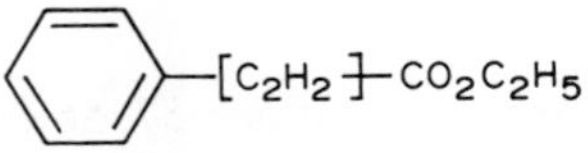

(5.73)

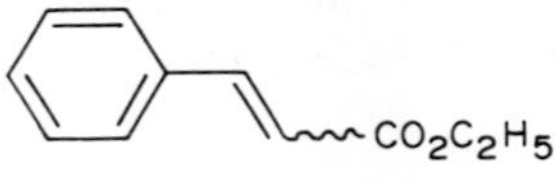

(5.74) 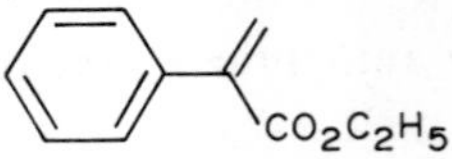   (5.75)

BIBLIOGRAPHY

Beynon, J. H., Saunders, R. A. and Williams, A. E. (1968), *The Mass Spectra of Organic Molecules*, Elsevier, Amsterdam.
Hill, H. C. (revised by. Loudon, A. G.) (1972), *Introduction to Mass Spectrometry*, 2nd ed., Heyden, London.
McFadden, W. H. (1973), *Techinques of Combined Gas Chromatography/Mass Spectrometry: Applications to Organic Chemistry*, Wiley–Interscience, New York.
McLafferty, F. W. (1973), *Interpretation of Mass Spectra*, 2nd ed., Benjamin, New York.
Morris, H. R. (Ed.) (1981), *Soft Ionisation Biological Mass Spectrometry*, Heyden, London.
Porter, Q. N. and Baldas, J. (1971), *Mass Spectrometry of Heterocyclic Compounds*, Wiley–Interscience, New York.
Rose, M. E. and Johnstone, R. A. W. (1982), *Mass Spectrometry for Chemists and Biochemists*, Cambridge University Press, Cambridge.
Waller, G. R. (Ed.) (1972), *Biochemical Applications of Mass Spectrometry*, Wiley–Interscience, New York.
Waller, G. R. and Dermer, O. C. (Eds) (1980), *Biochemical Applications of Mass Spectrometry, First Supplementary Volume*, Wiley–Interscience, New York.
Williams, D. H. and Howe, I. (1972), *Principles of Organic Mass Spectrometry*, McGraw-Hill, New York.

Compilations and Tables

Beynon, J. H., Saunders, R. A. and Williams, A. E. (1965), *Tables of Metastable Transitions for Use in Mass Spectrometry*, Elsevier, Amsterdam.

Beynon, J. H. and Williams, A. E. (1963), *Mass and Abundance Tables for Use in Mass Spectrometry*, Elsevier, Amsterdam.

Cornu, A. and Massot, R. (1975), *Compilation of Mass Spectral Data*, 2nd ed., Heyden, London.

Lederberg, J. (1964), *Computation of Molecular Formulas for Mass Spectrometry*, Holden-Day, San Francisco.

McLafferty, F. W. (1963), *Mass Spectral Correlations*, American Chemical Society, Washington, DC.

Pretch, E., Clerc, T., Seibl, J. and Simon, W. (1983), *Spectral Data for Structure Determination of Organic Compounds*, Springer-Verlag, Berlin, Heidelberg, New York and Tokyo.

CHAPTER 6

Problems

6.1 NOTES ON PROBLEMS

The problems 1–27 are arranged in approximate order of difficulty and include ultraviolet data, where relevant, infrared, mass, proton and carbon NMR spectra, together with molecular formulae or elemental analyses. Problems 27–45 contain selected data in tabulated form, in addition to chemical information.

6.1.1 Mass spectra

All mass spectra were recorded at 70 eV under EI conditions, unless indicated otherwise. The use of low eV (*ca* 15 eV) results in less fragmentation and a relatively stronger molecular ion. Where CI conditions were used, the reagent gas was isobutane. This causes protonation of the molecular species, which will therefore appear as a pseudo-molecular ion at $m/z\ M + 1^+$. This phenomenon is also observed where FAB conditions are employed using a glycerol matrix. These last two techniques cause much less fragmentation than EI and, where fragmentation is observed, the daughter species may or may not be protonated.

6.1.2 Molecular formulae

These are readily determined from the elemental analyses together with the mass spectral data, provided that the molecular ion or pseudo-molecular ion can be located, in which case

$$\frac{M \times \%X}{100 \times \text{RAM X}} = \text{number of X atoms}$$

(RAM = Relative Atomic Mass)

Oxygen atoms, as usual, can be determined by difference.

6.1.3 Ultraviolet spectra

These were usually obtained on solutions in 95% ethanol using 1 cm path-length silica cells. Where no data are provided in problems 1–27, it may be assumed that there is no significant absorption above 200 nm.

6.1.4 Infrared spectra

The sample conditions used to measure the spectrum are described in each case. Note that Nujol gives rise to bands at 3000–2850, 1468, 1379 and 720 (weak) cm^{-1}.

6.1.5 NMR spectra

Unless it is stated otherwise, both proton and carbon NMR spectra were determined in deuterochloroform at 60 and 22.5 MHz, respectively, with TMS as internal standard. The multiplicities of the ^{13}C signals are tabulated as they would appear in an off-resonance spectrum, that is, as quartets (q), triplets (t), doublets (d) and non proton-bearing (s).

The solvents used often appear in the ^{13}C NMR spectra as relatively low-intensity signals: CCl$_4$(96.1, s), CDCl$_3$(77.0, t), CD$_2$Cl$_2$(53.4, p), DMSO-d_6 and acetone-d_6 [29.8 and 39.6 (hept)]. These may be distinguished from solute resonances by the fact that they do not change in the off-resonance and are absent from the DEPT spectra.

Deuterium has $I = 1$ and hence three possible spin states $(+1, 0, -1)$, so deuterochloroform produces a triplet of ratio 1:1:1 in its carbon spectrum. Clearly the $n + 1$ rule only applies to nuclei of spin $\frac{1}{2}$. The multiplicity of the carbon resonance here is given by $2n + 1$, where n is the number of equivalent, adjacent deuterium atoms. Deuterium relaxes the carbon spin states less efficiently than hydrogen, hence the relative weakness of the solvent signals.

Finally, beware of fortuitous coincidences of resonances in the carbon NMR spectra. If they correspond to identical carbon types, they may be difficult to detect.

6.2 PROBLEMS

Problem 1

C, 80.6; H, 7.5%.
$\lambda_{max}(\varepsilon)$ 242 (12 600), 279 (1050), 318 (60) nm.

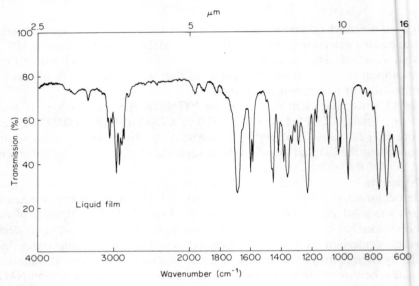

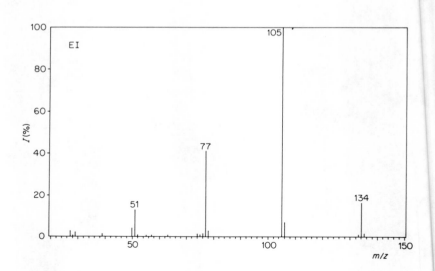

PROBLEMS

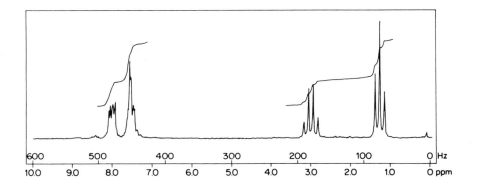

^{13}C data
1   8.2(q)
2   31.6(t)
3 127.9(d)
4 128.6(d)
5 132.8(d)
6 137.1(s)
7 200.2(s)

Problem 2

C, 66.6; H, 11.2%.
$\lambda_{max}(\varepsilon)$ 290(18) nm.

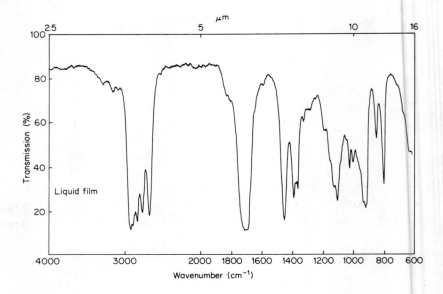

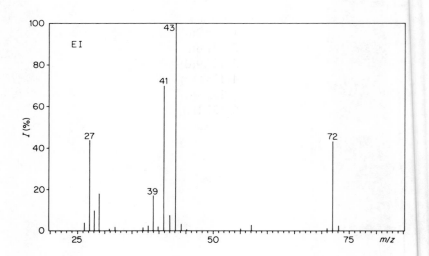

PROBLEMS

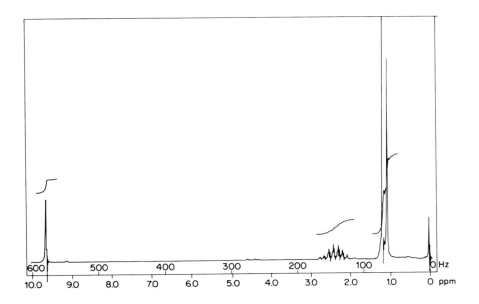

^{13}C data
1 15.3(q)
2 41.5(d)
3 205.0(d)

Problem 3

C, 56.7; H, 4.8; N, 6.6; Cl, 16.75%.
$\lambda_{max}(\varepsilon)$ 236 (9800) nm.

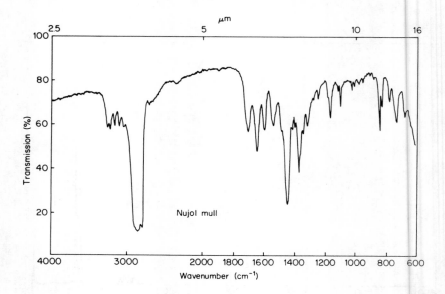

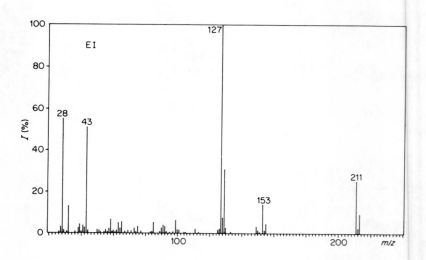

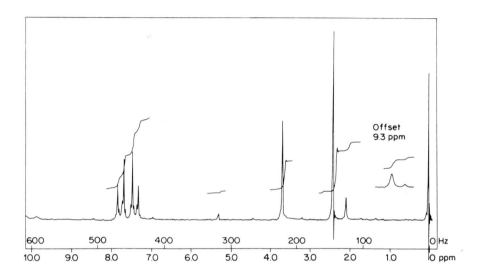

^{13}C (CDCl$_3$/DMSO-d_6)
1 30.1(q)
2 52.2(t)
3 120.8(d)
4 129.0(s)
5 129.2(d)
6 137.8(s)
7 165.2(s)
8 202.4(s)

Problem 4

C, 63.45; H, 5.8%.
λ_{max} (ε) 212 (10890), 237 (12690), 342 (2580) nm.

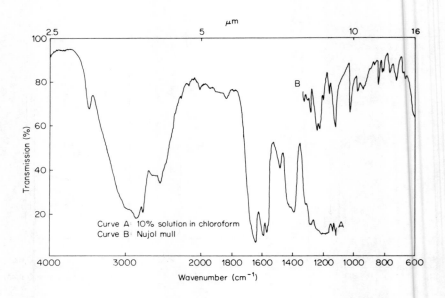

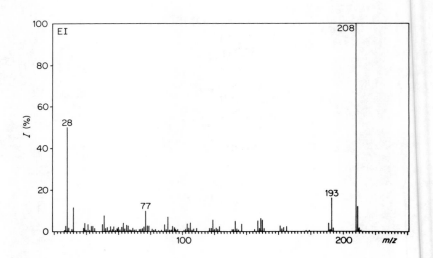

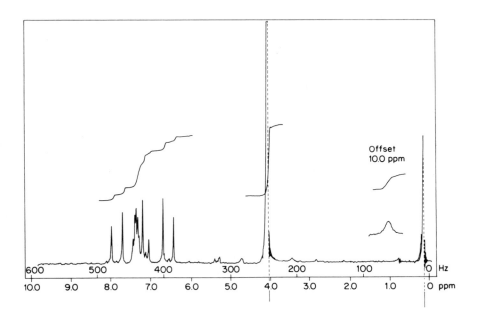

^{13}C data
1 55.4(q)
2 109.4(d)
3 111.0(d)
4 121.8(d)
5 128.0(s)
6 140.9(d)
7 147.8(s)
8 149.0(s)
9 175.9(s)

Problem 5

C, 64.0; H, 4.0%.
$\lambda_{max}(\varepsilon)$ 210 (21 500), 236 (20 000), 276 (18 000), 320 (85) nm.

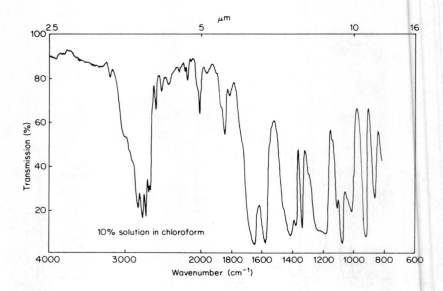

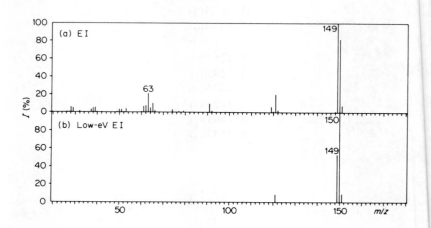

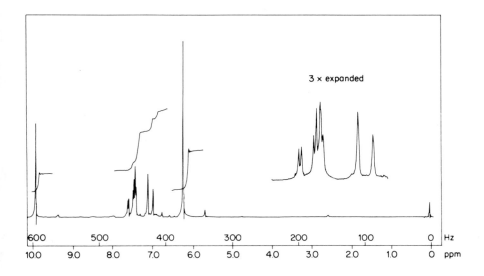

^{13}C data
1 102.4(t)
2 107.0(d)
3 108.4(d)
4 128.6(d)
5 132.3(s)
6 150.0(s)
7 153.2(s)
8 190.1(d)

Problem 6

C, 28.8; H, 4.2; Br, 47.8%.
$\lambda_{max}(\varepsilon)$ 203(35) nm.

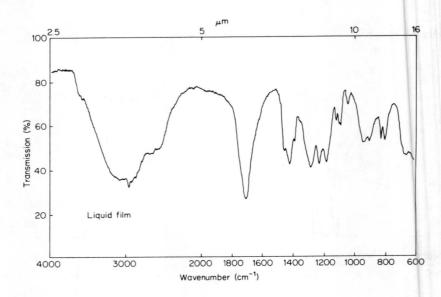

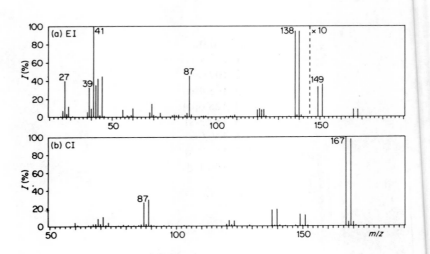

PROBLEMS

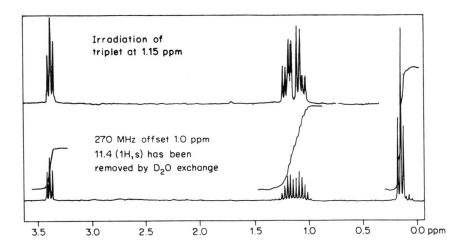

Irradiation of triplet at 1.15 ppm

270 MHz offset 1.0 ppm
11.4 (1H,s) has been removed by D$_2$O exchange

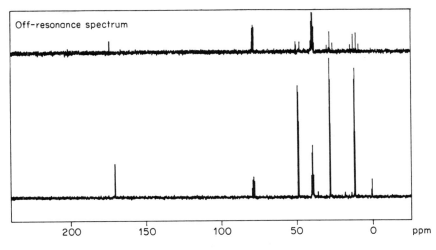

Off-resonance spectrum

Problem 7

C, 57.2; H, 7.8; N, 10.3; Cl, 13.0%.
$\lambda_{max}(\varepsilon)$ 222 (4400), 299 (8500) nm.

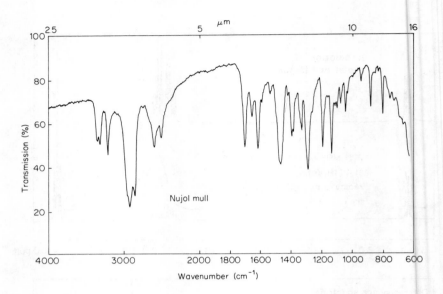

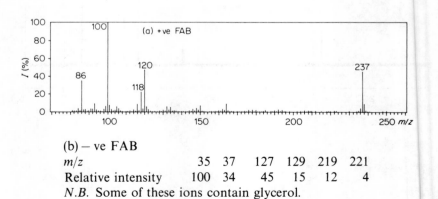

(b) − ve FAB

m/z	35	37	127	129	219	221
Relative intensity	100	34	45	15	12	4

N.B. Some of these ions contain glycerol.

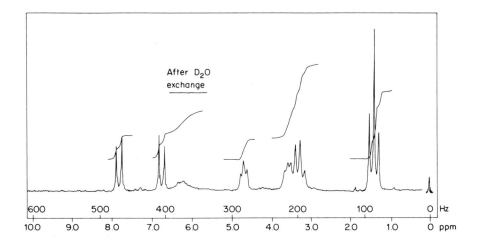

^{13}C data
1 10.9(q)
2 51.0(t)
3 53.0(t)
4 61.6(t)
5 117.0(d)
6 119.6(s)
7 134.6(d)
8 156.1(s)
9 170.3(s)

Problem 8

C, 57.4; H, 8.6%.
$\lambda_{max}(\varepsilon)$ 265 (60) nm.

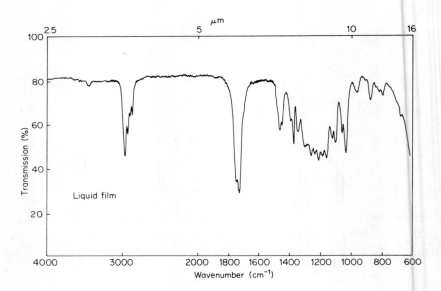

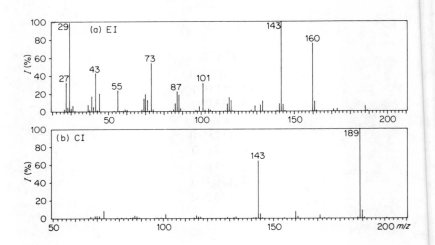

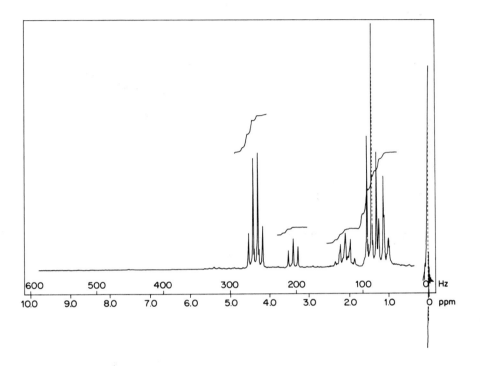

^{13}C data
1 12.0(q)
2 14.3(q)
3 22.7(t)
4 54.0(d)
5 61.3(t)
6 170.0(s)

Problem 9

C, 55.3; H, 5.3; N, 18.4%.
$\lambda_{max}(\varepsilon)$ 221 (44 600), 243 (36 870), 270 (19 095), 322 (14 550) nm.

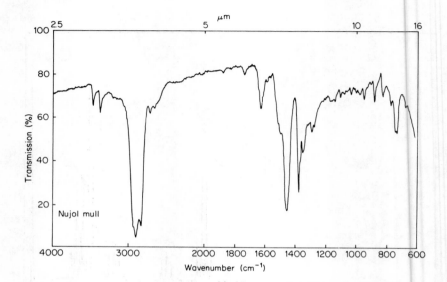

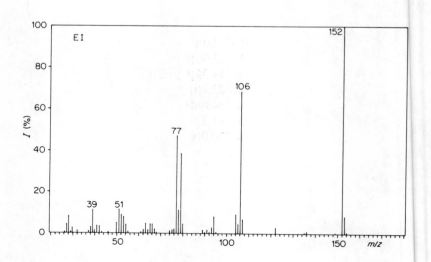

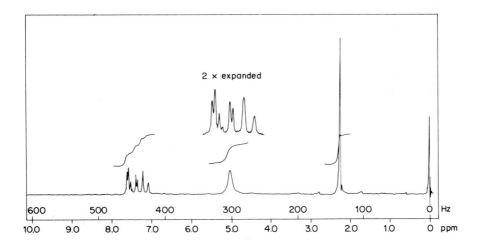

¹³C data
1 17.3(q)
2 112.1(d)
3 120.3(s)
4 123.5(d)
5 125.8(d)
6 136.2(s)
7 153.0(s)

Problem 10

C, 16.1; H, 2.0; Cl, 71.2%.

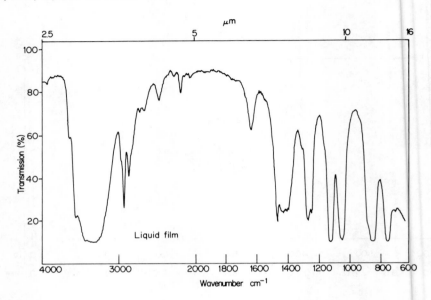

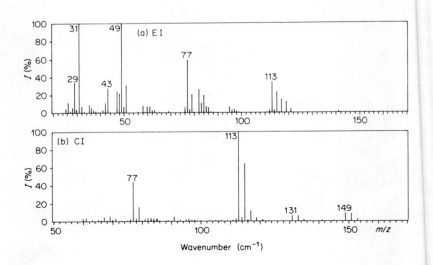

PROBLEMS

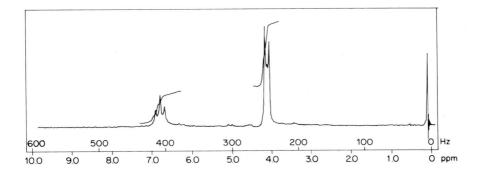

^{13}C (DMSO-d_6)
1 75.5(t)
2 97.8(s)

Problem 11

C, 53.3; H, 4.5; N, 15.55%.
$\lambda_{max}(\varepsilon)$ 237 (36 670), sh. 273 (10 870), 322 (6560) nm.

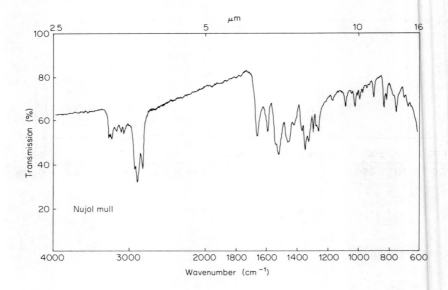

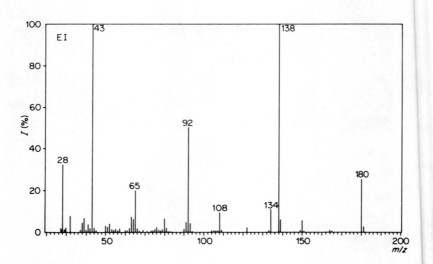

PROBLEMS

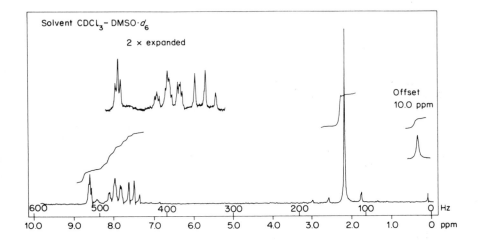

^{13}C data
1 24.1(q)
2 124.0(d)
3 125.1(d)
4 133.6(s)
5 134.6(d)
6 140.1(s)
7 169.1(s)

Problem 12

C, 71.0; H, 7.95.
$\lambda_{max}(\varepsilon)$ 220 (8000), 260 (20000), 273 (2200), 280 (2500) nm.

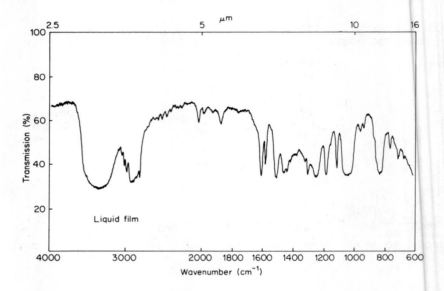

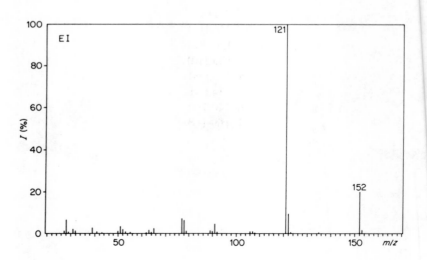

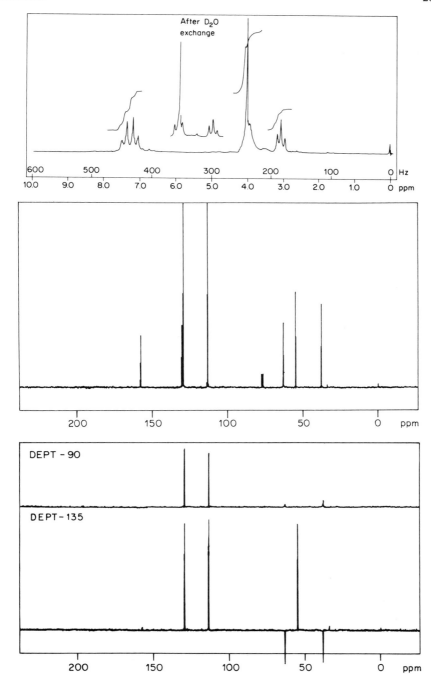

Problem 13

C, 58.5; H, 3.4; N, 6.8%.
$\lambda_{max}(\varepsilon)$ 217 (38 000), 273 (1800), 296 (1800) nm.

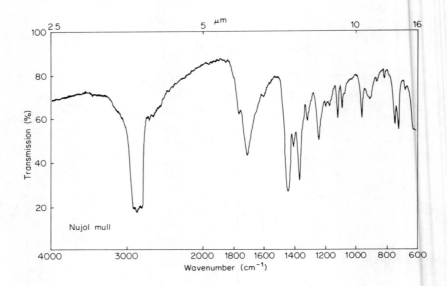

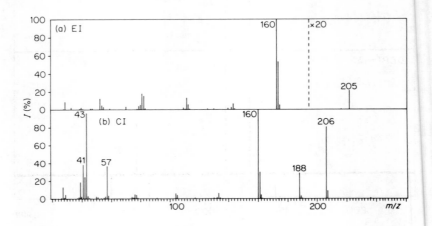

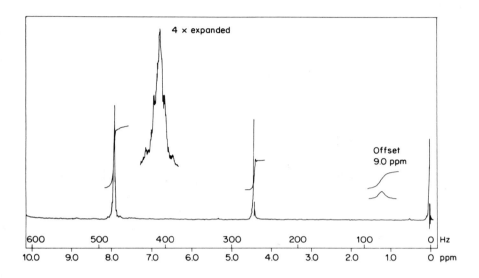

^{13}C data
1 39.0(t)
2 123.4(d)
3 131.5(s)
4 134.5(d)
5 168.2(s)
6 169.1(s)

Problem 14

C, 70.0; H, 6.7; N, 23.3%.
$\lambda_{max}(\varepsilon)$ 210 (5100), 240 (3000) nm.

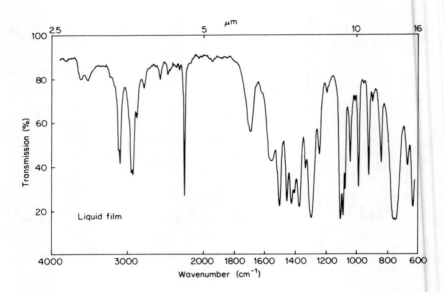

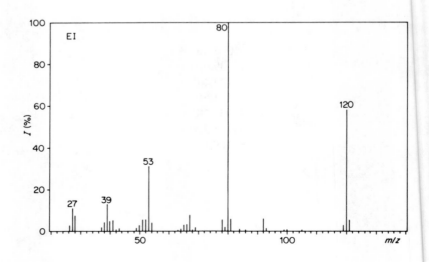

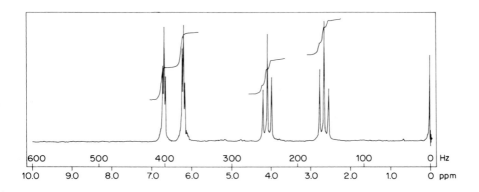

^{13}C data
1 20.3(t)
2 44.6(t)
3 108.9(d)
4 118.0(s)
5 120.5(d)

Problem 15

C, 58.8; H, 13.8; N, 27.4%.

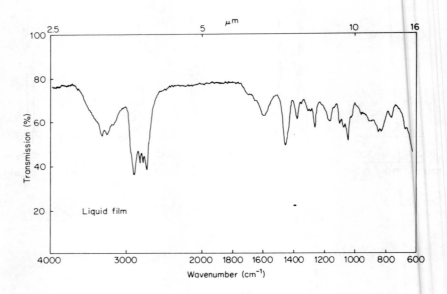

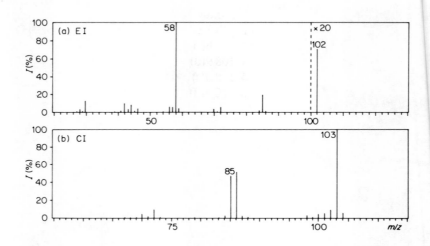

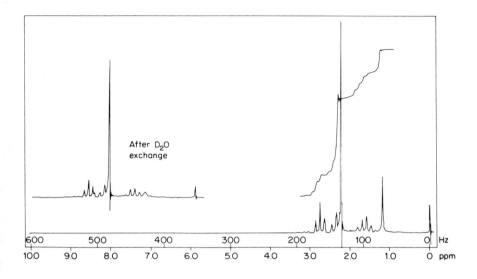

^{13}C data
1 31.6(t)
2 40.7(t)
3 45.7(q)
4 57.7(t)

Problem 16

C, 62.5; H, 4.2%.
$\lambda_{max}(\varepsilon)$ 219 (24 000), 278 (15 800) nm.

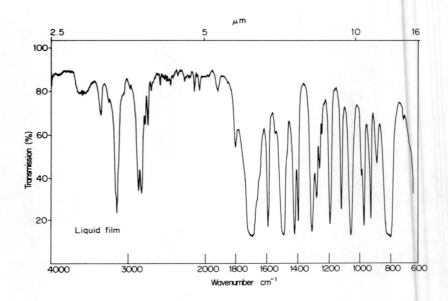

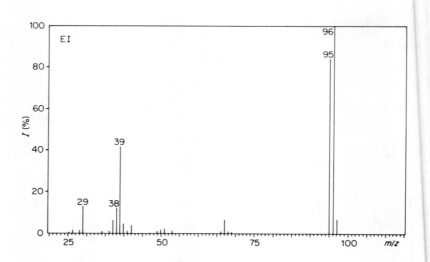

PROBLEMS

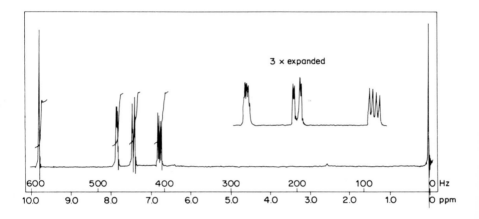

3 × expanded

^{13}C data
1 113.6(d)
2 122.4(d)
3 149.5(d)
4 154.3(s)
5 178.6(d)

Problem 17

C, 72.2; H, 5.3; N, 10.5%.
$\lambda_{max}(\varepsilon)$ 203 (6128), 246 (5250) nm.

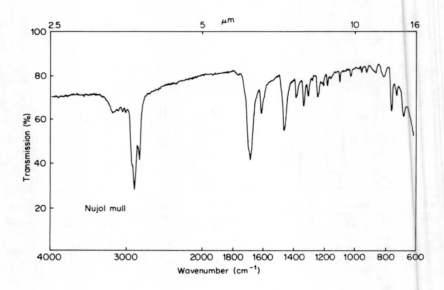

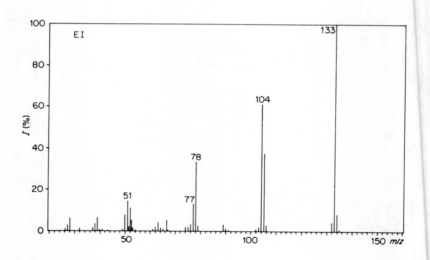

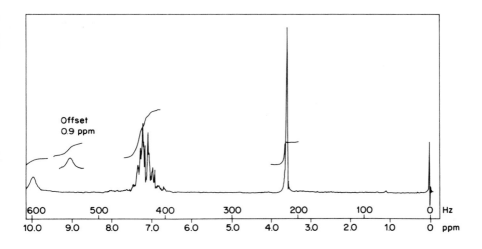

^{13}C (DMSO-d_6)
1 36.0(t)
2 109.3(d)
3 121.1(d)
4 124.4(d)
5 125.8(s)
6 127.6(d)
7 143.7(s)
8 176.6(s)

Problem 18

C, 63.7; H, 9.8; N, 12.4%.
$\lambda_{max}(\varepsilon)$ 243 (77) nm.

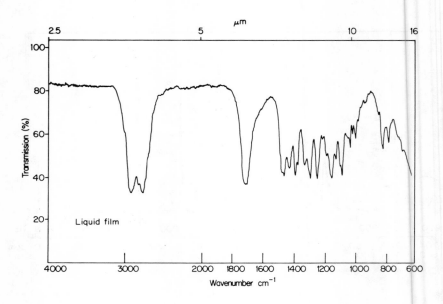

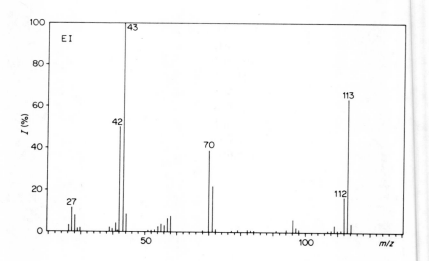

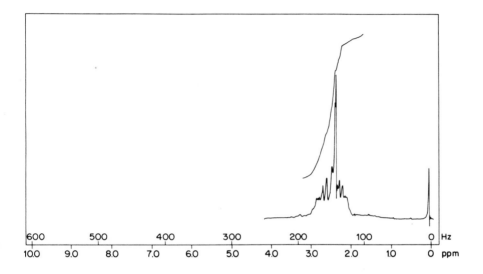

¹³C data
1 41.1(t)
2 45.2(q)
3 53.3(t)
4 207.7(s)

Problem 19

C, 68.0; H, 7.3%.
$\lambda_{max}(\varepsilon)$ 210 (8300), 263 (300) nm.

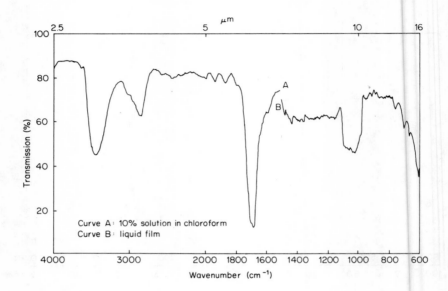

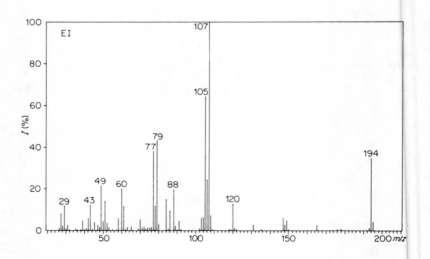

PROBLEMS 223

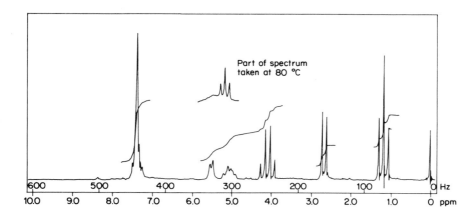

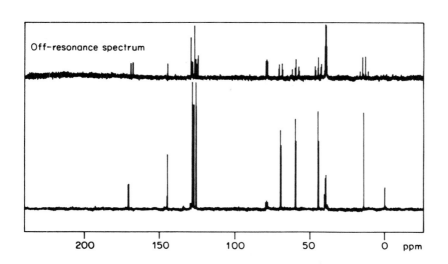

Problem 20

C, 59.0; H, 4.95; N, 22.9%;
$\lambda_{max}(\varepsilon)$ 213 (30 326) nm.

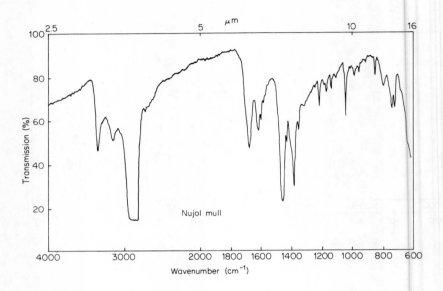

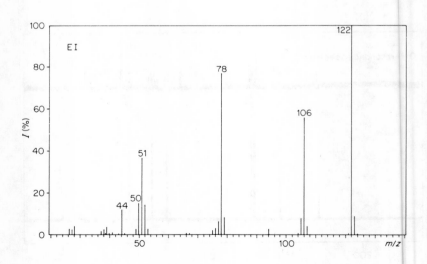

PROBLEMS

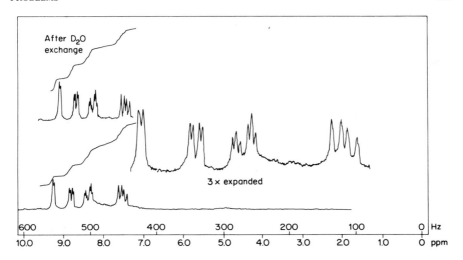

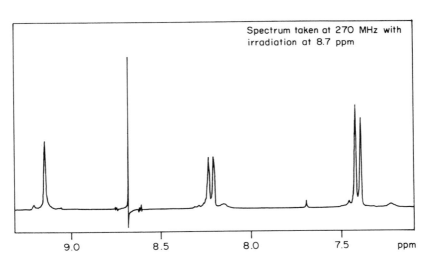

^{13}C (DMSO-d_6)
1 123.1(d)
2 129.6(s)
3 136.7(d)
4 148.5(d)
5 152.7(d)
6 170.8(s)

Problem 21

C, 41.2; H, 2.3; F, 10.85; Br, 45.65%.
$\lambda_{max}(\varepsilon)$ 226(6500), 260(1200), 269(1200) nm.

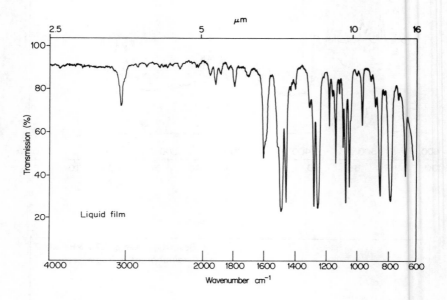

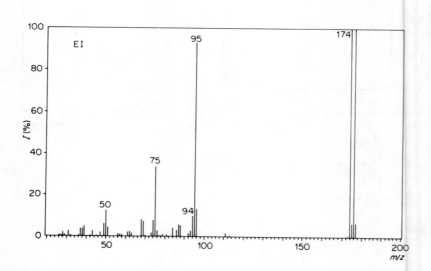

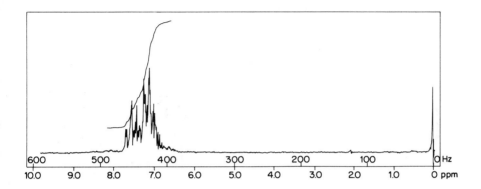

^{13}C data
1 109.5(d, ~ 20 Hz)
2 116.2(d, ~ 20 Hz)
3 124.5(d, ~ 5 Hz)
4 128.6(d, ~ 8 Hz)
5 133.8(d, ~ 8 Hz)
6 159.5(d, ~ 250 Hz)

Problem 22

C, 55.8; H, 7.0%.
$\lambda_{max}(\varepsilon)$ 268 (40) nm.

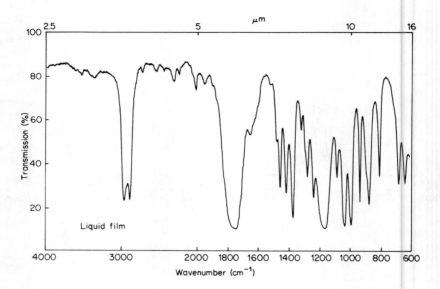

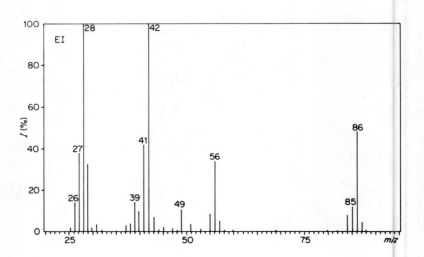

PROBLEMS

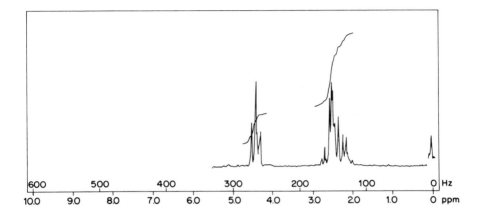

^{13}C data
1 22.3(t)
2 27.8(t)
3 68.8(t)
4 178.0(s)

Problem 23

C, 61.0; H, 11.9%.

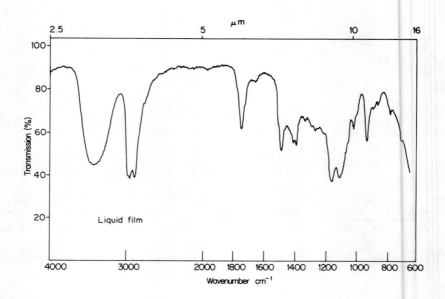

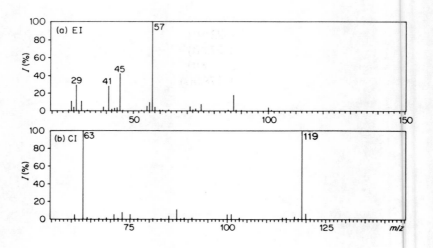

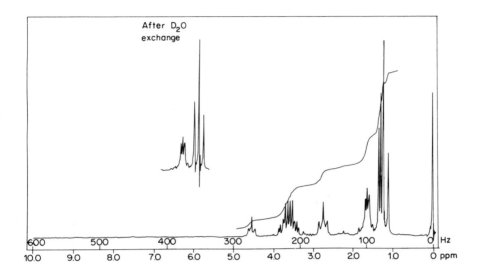

^{13}C data
1   14.5(q)
2   24.8(t)
3   32.2(t)
4   41.7(t)
5   57.5(t)
6 104.0(d)

Problem 24

C, 59.6; H, 11.9; N, 8.7%.

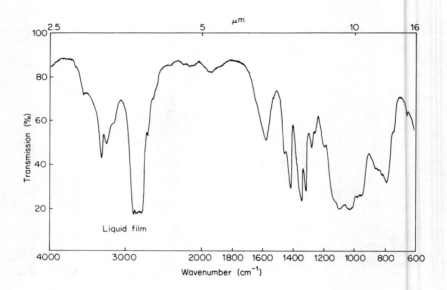

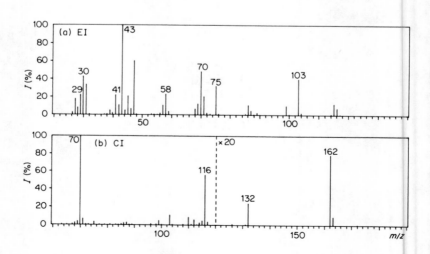

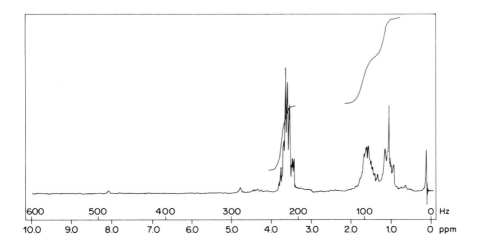

^{13}C data
1 13.7(q)
2 19.2(t)
3 31.7(t)
4 61.4(t)
5 71.1(t)
6 72.1(t)

Problem 25

C, 74.1; H, 7.9%.
$\lambda_{max}(\varepsilon)$ 213 (17 000), 263 (15 600), 303 (6600) nm.

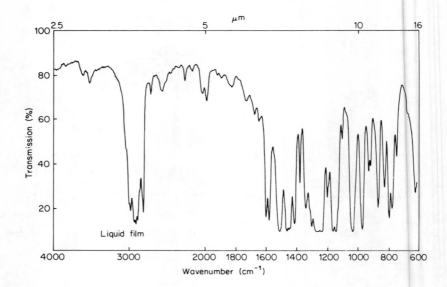

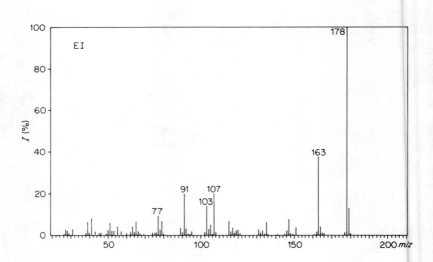

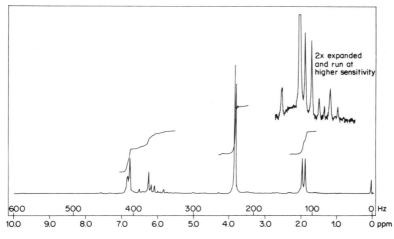

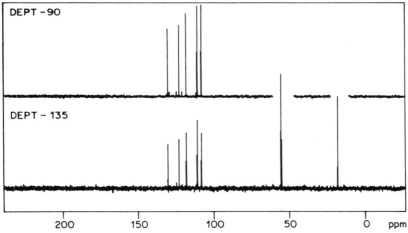

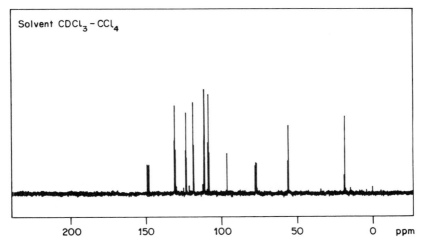

Problem 26

C, 70.6; H, 4.85; N, 7.5%.

$\lambda_{max}(\varepsilon)$ 218 (53 000), 266 (11 200), 330 (4000) nm.

The problem of orientation of substituents can be resolved by consideration of the following information. Double irradiation of the obscured doublet ($J = 8.0$ Hz centred at 7.75 ppm causes the doublet at 7.25 ppm to collapse to a singlet, and irradiation at 2.5 ppm in an nOe experiment caused enhancement of the 7.25 ppm doublet.

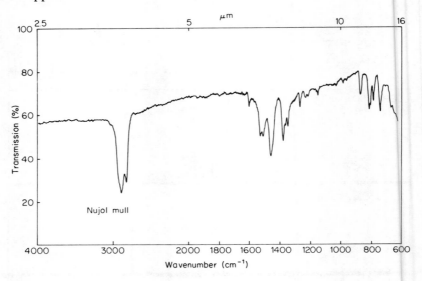

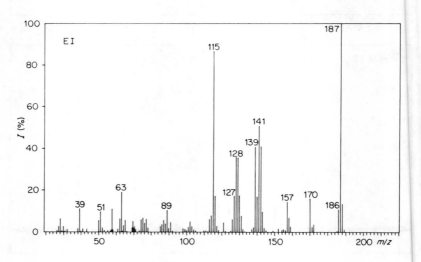

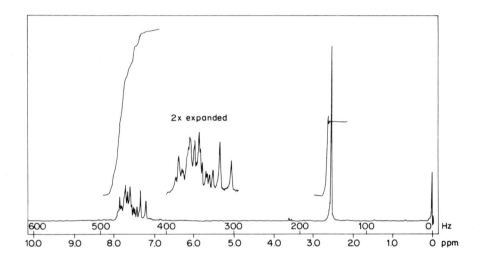

^{13}C data
1 17.6(q)
2 120.8(d)
3 121.0(d)
4 124.1(s)
5 126.2(d)
6 127.0(d)
7 127.5(s)
8 128.0(d)
9 131.8(s)
10 147.5(s)

Problem 27

C, 74.9; H, 12.6% ($C_8H_{16}O$).

This sample was labelled by the manufacturer '3,4-dimethylcyclohexanol' and described as 'pure.' Interpret the IR and proton NMR spectra as far as possible and by examining the carbon spectrum, particularly in the 64–72 ppm expanded region, decide what are the most likely constituents of the sample.

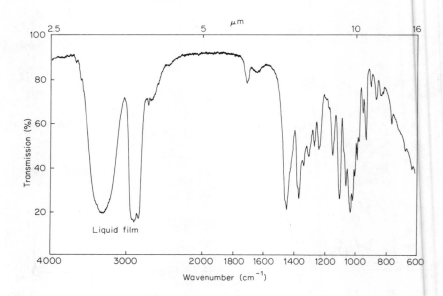

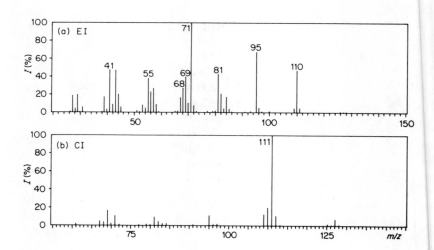

PROBLEMS

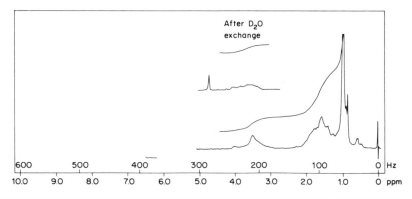

Solvent $CDCl_3 - CCl_4$

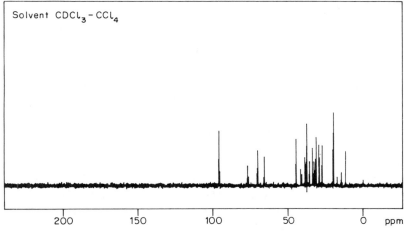

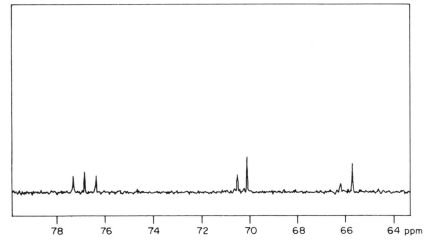

Problem 28

Oxidation of 5-methoxy-2-nitrotoluene with potassium permanganate yields A, m/z 197 (20%, $M^{+\cdot}$), 153 (100%), ν_{max} 3550, 3300–2600, 1680, 1540, 1310 cm^{-1}.

A is reduced by hydrogenation over a platinum catalyst to B, m/z 167 (25%, $M^{+\cdot}$), ν_{max} 3500, 3360–2650, 1680 cm^{-1}.

Reaction of B with monochloroacetic acid in the presence of aqueous alkali affords C, $C_{10}H_{11}NO_5$, which on treatment with boiling acetic anhydride containing sodium carbonate, followed by acidification, gives D, $C_{13}H_{13}NO_4$. Physical data for D: ν_{max} 1740, 1700, 1610 cm^{-1}; δ_H 2.30 (s, 3H), 2.45 (s, 3H), 3.95 (s, 3H), 6.95 (d, $J = 2$ Hz, 1H), 7.1 (dd, $J = 9$ and 2 Hz, 1H), 8.05 (d, $J = 9$ Hz, 1H), 8.15 (s, 1H).

Identify compounds A–D.

Problem 29

When 3,4-methylenedioxy-β-nitrostyrene is heated with furan and a catalytic amount of zinc iodide, compound A is produced. Physical data for A: $C_{13}H_{11}NO_5$, ν_{max} 3070–3020, 2780, 1600, 1550, 1480, 1370 cm^{-1}; δ_H 4.90 (m, 3H), 6.00 (s, 2H), 6.15 (d, $J = 5$ Hz, 1H), 6.35 (dd, $J = 5$ and 5 Hz, 1H) 6.78 (bs, 3H), 7.36 (d, $J = 5$ Hz, 1H); δ_C 43.3 (d), 78.2 (t), 101.4 (t), 107.3 (d), 108.2 (d), 108.7 (d), 110.5 (d), 121.4 (d), 130.7 (s), 142.6 (d), 147.5 (s), 148.2 (s), 152.2 (s).

When reacted, first with sodium hydride, and then with bromine, A yields the diastereomers B as a mixture: m/z 341 (50%), 339 (50%, $M^{+\cdot}$), 294 (30%), 292 (30%), 201 (100%); δ_H 4.21 (d, $J = 6$ Hz, 1H), 4.85 (d, $J = 6$ Hz, 1H), 5.95 (s, 2H), 6.20 (d, $J = 5$ Hz, 1H), 6.35 (dd, $J = 5$ and 5 Hz, 6.8 (bs, 3H), 7.35 (d, $J = 5$ Hz, 1H).

Identify compounds A and B.

Problem 30

Nicotinoyl chloride and indol-3-ylmagnesium bromide react together to give A, m/z 222 (20%, $M^{+\cdot}$), 144 (100%); λ_{max} (ε) 258 (10 000) 265 (10 500), 317 (9320) nm; ν_{max} 3100, 1600 (strong), 1580, 1230 cm^{-1}; δ_H 7.5 (bs, 3H), 7.60 (s, 1H), 8.15 (d, $J = 9$ Hz, 1H), 8.2 (m, 2H), 8.86 (d, $J = 8$ Hz, 1H), 9.05 (s, 1H), 12.21 (s, 1H, exchanged on shaking sample with D_2O).

A, when heated with sodium borohydride in aqueous ethanol, affords B, m/z 208 (15%, $M^{+\cdot}$), 130 (100%), 117 (10%); λ_{max} (ε) 270 (8500), 282 (7400), 291 (6400) nm; δ_H 4.12 (s, 2H), 6.9 (m, 3H), 7.05 (s, 1H), 7.4 (d, $J = 9$ Hz), 7.66 (m, 2H), 8.45 (d, $J = 8$ Hz, 1H), 8.64 (s, 1H), 10.45 (s, 1H, exchanged on shaking the sample with D_2O).

Identify compounds A and B.

PROBLEMS 241

Problem 31

3,4-Dimethoxyphenylacetic acid in glacial acetic acid, when heated with a mixture of hydrochloric acid and formaldehyde, produces a colourless solid A. Physical data for A: C 63.45, H5.8%; m/z 208(100%, $M^{+\cdot}$), 164(77%); ν_{max} 1730, 1605 cm^{-1}; δ_H 3.65(s, 2H), 3.90(s, 6H), 5.25(s, 2H), 6.70(s, 1H), 6.75(s, 1H); δ_C 35.5(t), 55.7(q), 69.3(t), 109.1(d), 110.8(d), 123.6(s), 124.1(s), 147.8(s), 149.0(s), 171.0(s).

When A is heated with 3,4-dimethoxybenzaldehyde and a base (pyrrolidine) a yellow solid B is formed. Physical data for B: C 67.4, H 5.7%; λ_{max} 246, 362, 480 nm; ν_{max} 1720, 1625 cm^{-1}; δ_H 3.58(s, 2 × 3H), 3.75(s, 3H), 3.91(s, 3H), 5.29(s, 2H), 6.29(s, 1H), 6.82(d, $J = 8$ Hz, 1H), 7.01(s, 1H), 7.11(d, $J = 2$ Hz, 1H), 7.15(dd, $J = 8$ and 2 Hz, 1H), 7.71(s. 1H); δ_c 55.2, 55.3, 55.4, 55.6(4 × q), 68.4(t), 108.9, 109.9, 111.6, 113.0(4 × d), 122.1, 122.7(2 × d), 123.2, 126.1, 126.6, 135.4, 147.8, 148.4, 149.1, 150.1, 167.8 (9 × s).

Identify compounds A, and B. How would you establish the stereochemistry of B?

Problem 32

Compound A, $C_{13}H_8O$, m/z 180(100%, $M^{+\cdot}$), 152(20%, 'metastable' 128.4); ν_{max} 1700, 1590, 1290, 1180, 1140, 1090, 900 cm^{-1}; δ_H 7.0–7.3 (m, 6H), 7.55(dd, $J = 9{,}2$ Hz, 2H), when heated with hydroxylamine gives B, $C_{13}H_9NO$. C is obtained from B by heating with polyphosphoric acid. Physical data for C: m/z 195(100%, $M^{+\cdot}$), 167(10%, M* 143.0); ν_{max} 3220–3100, 1635, 775, 750 cm^{-1}; δ_H 7.2–8.0(m, 5H), 8.4–8.6(m, 3H), 11.8(bs, 1H, slowly exchanged when the sample is shaken with D_2O).

Identify compounds A, B and C.

Problem 33

A is formed by the reaction of 3,4-dimethoxyphenylacetic acid and N, N'-carbonyldiimidazole, followed by treatment of the product with lithium aluminium hydride. Physical data for A: ν_{max} 1720, 1604, 1590 cm^{-1}; δ_H 3.64(s, 4H), 3.82(bs, 12H), 6.75(dd, $J = 8$ and 2 Hz, 2H), 6.85(d, $J = 2$ Hz, 2H), 6.90(d, $J = 8$ Hz, 2H).

A, when electrolysed at platinum anode or treated with vanadium oxytrifluoride (a one-electron oxidant), yields B, $C_{19}H_{20}O_5$; m/z 328(25%, $M^{+\cdot}$), 151(100%), ν_{max} 1705, 1600, 1585 cm^{-1}; δ_H 3.52(s, 4H), 4.03(bs, 12H), 6.80(s, 2H), 7.10(s, 2H).

Identify compounds A and B.

Problem 34

A, $C_6H_{10}O_2$, λ_{max} 260(ε 19) nm; ν_{max} 1710, 1360, 1158 cm^{-1}; δ_H 2.15(s, 6H), 2.74(s, 4H), when heated with aqueous sodium hydroxide affords B m/z 96(100%, $M^{+\cdot}$),

81 (45%), 67 (57%), 53 (50%); λ_{max} 228 (ε 15 000) nm; λ_{max} 1700, 1610, 1180, 975, 840 cm^{-1}; δ_H 2.18(d, $J = 1.5$ Hz, 3H), 2.28(t, $J = 6$ Hz, 2H), 2.34(t, $J = 6$ Hz, 2H), 5.9(q, $J = 1.5$ Hz, 1H); δ_C 19.1(q), 33.1, 35.8 (2 × t), 130.9(d), 177.8(s), 209.2(s).
Identify compounds A and B.

Problem 35

Furan, when heated with methyl acrylate and a trace of Lewis acid, ZnI_2, gives a mixture of two adducts A and B. A, when treated with osmium tetraoxide and hydrogen peroxide gives C, $C_8H_{12}O_5$, m/z (CI, isobutane) 189(100%, $M + 1^+$), 171(35%), 157(35%); v_{max} 3350, 1740 cm^{-1}; δ_H 1.58 (dd, $J = 12$ and 8 Hz, 1H), 2.10(qd, $J = 12$ and 4 Hz, 1H), 2.52(dd, $J = 8$ and 4 Hz, 1H), 3.30(s, 2H, exchanged when the sample is shaken with D_2O), 3.70(s, 3H), 3.85(s, 2H), 4.38(d, $J = 4$ Hz, 1H), 4.54(s, 1H).

B affords D, an isomer of C, which has similar infrared and mass spectra to C and in the ^1H NMR spectrum exhibits δ 1.60(m, 1H), 2.10(m, 1H), 3.1(m, 1H), 3.25(s, 2H, exchanged when the sample is shaken with D_2O), 3.68(s, 3H), 3.85 (s, 2H).

Identify compounds A–D (molecular models will prove helpful in deducing the dihedral angles between the C—H bonds in these products so that the Karplus equation, pp. 128–129, can then be used to predict H–H coupling constants).

Problem 36

N-Benzyloxycarbonylglycylglycine methyl ester ($PhCH_2OCONHCH_2CONH \cdot CH_2CO_2CH_3$) and phosphorus pentasulphide (P_2S_5) were heated in toluene for 1 h at 90 °C, yielding a white solid A.

A showed IR v_{max} 3400, 3300, 1745, 1715 and 1230 cm^{-1}. The mass spectrum contained peaks at m/z 296 ($M^{+\cdot}$28), 237(6), 188(8), 135(34) and 91 (100). The peak at m/z 296 was accompanied by isotope satellites, such that m/z 296:297:298 were in the ratio 100:16.1:6.52. Accurate mass determination of m/z 296 gave the value 296.0829. The ^{13}C NMR spectrum consisted of 11 lines, δ 46.7(t), 51.9(q), 52.5(t), 67.4(t), 128.0(d), 128.2(d), 128.6(d), 136.1(s), 156.8(s), 168.8(s) and 200.7(s). There are coincidences in this spectrum, the intensity of the peaks at δ 128.0 and 128.6 being approximately twice that of the peak at δ 128.2.

Identify compound A.

Problem 37

When fructose is heated in butan-2-ol in the presence of a strongly acidic resin, two compounds are produced. Compound A has v_{max} 1740 and 1720 cm^{-1}; m/z 99 and 43 (no $M^{+\cdot}$ is detected under EI conditions); δ_H 0.9 (t, 3H, $J = 7$Hz), 1.2(d, 3H, $J = 7$Hz), 1.55 (m, 2H), 2.2(s, 3H), 2.2–2.9(m, 4H, symm.), 4.85(sex, 1H, $J = 7$Hz). Compound B has v_{max} 1680 cm^{-1}; m/z 109 and 43 (no $M^{+\cdot}$ is detected

under EI conditions); δ_H 0.9 (t, 3H, $J = 7$ Hz), 1.2(d, 3H, $J = 7$ Hz), 1.55(m, 2H), 3.5 (sex, 1H, J = 7 Hz), 4.6 (s, 2H), 6.25 and 7.25 (d, 1H, $J = 3.5$ Hz), 9.55 (s, 1H); δ_C 9.5(q), 19.1(q), 29.1(t), 62.7(t), 77.2(d), 110.7(d), 122.1(d), 152.7(s), 159.6(s), 177.5(d).
Identify compounds A and B.

Problem 38

The methyl ester of serine, HOCH$_2$CH(NH$_2$)COOMe, if treated successively with Ph$_3$CCl–base, p-toluenesulphonyl chloride–pyridine and then the base triethylamine, yields compound A. Physical data for A: v_{max} 1750 cm^{-1}; m/z (CI) 344(M + 1)$^+$, 285, 243; δ_H 1.4 (dd, 1H, $J = 1.5$ and 6.0 Hz); 1.9(dd, 1H, $J = 2.5$ and 6.0 Hz); 2.3 (dd, 1H, $J = 1.5$ and 2.5 Hz); 7.2–7.5 (m, 15H); δ_C 28.6(t), 31.7(d), 51.9(q), 74.4(s), 126.9(d), 128.2(d), 129.3(d), 143.6(s), 171.8(s).
Identify compound A.

Problem 39

N-(3,4-Methylenedioxybenzyl)aminoacetaldehyde dimethylacetal, (CH$_2$O$_2$)-C$_6$H$_3$CH$_2$NHCH$_2$CH(OMe)$_2$, acts as a nucleophile in a Michael reaction with methyl vinyl ketone. The addition product, on treatment with aqueous HCl, undergoes a series of acid-catalysed steps leading to compound A, which has the following spectral characteristics: v_{max} 2975, 2945, 2920, 2880, 1720 cm^{-1}; m/z 245 (M$^+$$^\cdot$), 148 (100%), 147, 97; δ_H 2.3–2.8 (m, 8H), 3.3(m. 1H), 3.35 and 3.65 (d, 1H, $J = 15$ Hz), 5.3 (s, 2H), 6.5 and 6.6 (s, 1H); C 68.4, H 6.0, N 5.9%.
Identify compound A. Hint: pay attention to the retro-Diels–Alder process in the mass spectrum.

Problem 40

The self-condensation reaction of propanone gives a compound A, $\lambda_{max}(\varepsilon)$ 237 (12000) nm.
Reduction of A using NaBH$_4$, followed by acidification of the reaction mixture, gives several products. A minor one is compound B, which has the following spectral properties: v_{max} 3550–3200, 1635, 889 cm^{-1}; δ_H 1.3 (d, 3H, $J = 7.5$ Hz), 1.8 (t, 3H, $J = 1.5$ Hz), 2.2 (d, 2H, $J = 7.5$ Hz), 2.6 (broad s, 1H), 4.0 (sex, 1H, $J = 7.5$ Hz), 4.8 (ten lines observed, 2H, ca 1.5 Hz spacing).
Identify compounds A and B.

Problem 41

2-Trimethylsilyloxyfuran undergoes an SnCl$_4$-catalysed reaction with undecanal to give compound A. Physical data for A: C 71.4, H 10.2%; v_{max} (CHCl$_3$) 3570–

3250, 2960, 2840, 1740, 1600, 1250 cm^{-1}; m/z (CI) 255 (M + I$^+$ 100%), 237 (40%), 97, 85, 84 (56%); δ_H (270 MHz) 0.88 (d, 3H, J = 7.5 Hz), 1.20–1.40 (16H), 1.6 (br, m, 2H), 1.96 (br, s removed by D$_2$O, 1H), 3.76 (dd, 1H, J = 3.0, 4.5 and 8.0 Hz), 4.99 (dd, 1H, J = 1.5, 1.5 and 4.5 Hz), 6.21 (dd, 1H, J = 1.5 and 6.0 Hz), 7.45 (dd, 1H, J = 1.5 and 6.0 Hz); δ_C 14.1(q), 22.6(t), 25.6(t), 29.3(t), 29.4(t), 29.45(t), 29.5(t), 29.6(t), 31.9(t), 33.2(t), 71.5(d), 86.2(d), 122.7(d), 153.6(d), 173.3(s).

Identify compound A.

Problem 42

Vinyl acetate and chlorosulphonyl isocyanate, ClSO$_2$N=C=O, react together in a 2 + 2 electrocyclic reaction under slightly alkaline, aqueous conditions to give compound A, v_{max} (CHCl$_3$) 3350, 1790, 1735 cm^{-1}, N-Substitution of A by benzyl glyoxalate, OHCCOOCH$_2$Ph, gives B, C 57.34, H 5.12, N, 4.81%.

When B is treated with thionyl chloride and the product C is dehalogenated by the action of tributyltin hydride, compound D is produced. Physical data for D: C 60.39, H 5.50, N 5.06%; v_{max} 1790, 1740, 1735 cm^{-1}; δ_H 2.1 (s, 3H), 3.1 (m, 2H), 4.05 and 4.50 (d, 1H, J = 16 Hz), 5.10 (s, 2H), 5.95 (dd, 1H, J = 2.0 and 1.0 Hz), 7.25 (s, 5H); δ_C 19.6(q), 42.0(t), 43.6(t), 66.3(t), 75.7(d), 127.4(d), 128.1(d), 128.5(d), 134.5(s), 164.7(s), 167.2(s), 170.5(s).

Identify compounds A and D.

6.3 ANSWERS TO PROBLEMS

1. 1-Phenylpropan-1-one (propiophenone).
2. 2-Methylpropanal.
3. N-(4-Chlorophenyl)-2-oxobutanamide (plussome enol form).
4. (E)-3-(3,4-Dimethoxyphenyl)prop-2-enoic acid.
5. 3,4-Methylenedioxybenzaldehyde.
6. (±)-2-Bromobutanoic acid.
7. 2-Diethylaminoethyl 4-aminobenzoate, HCl salt (procaine hydrochloride).
8. Diethyl ethylpropanedioate (diethyl ethylmalonate).
9. 2-Methyl-5-nitroaniline.
10. 2,2,2-Trichloroethanol.
11. N-(3-Nitrophenyl)ethanamide (3-nitroacetanilide).
12. 2-(4-Methoxyphenyl)ethanol.
13. Phthalimidoethanoic acid (N-phthaloylglycine).
14. N-(2-Cyanoethyl)pyrrole.
15. 3-Dimethylamino-1-aminopropane.
16. 2-Furancarboxaldehyde (furfural).
17. Indol-2(3H)-one (oxindole).
18. N-Methylpiperid-4-one.
19. (±)-Ethyl 3-hydroxy-3-phenylpropanoate.
20. 3-Pyridinecarboxamide (nicotinamide).
21. 1-Bromo-2-fluorobenzene.
22. 2(3H)-Dihydrofuranone (γ-butyrolactone).
23. 2-Butoxyethanol.
24. Diethyl acetal of 4-aminobutanal.
25. (E)-1-(3,4-Dimethoxyphenyl)propene.
26. 2-Methyl-1-nitronaphthalene.
27. 3,4-Dimethylcyclohexanol (mixture of diastereoisomers).
28. A: 2-Nitro-5-methoxybenzoic acid.
    B: 2-Amino-5-methoxybenzoic acid.
    C: N-(2-Carboxy-4-methoxyphenyl)glycine.
    D: 1-Acetyl-3-acetoxy-5-methoxyindole.
29. A: 2-(3,4-Methylenedioxyphenyl)-2-(furan-2-yl)nitroethane.
    B: 2-(3,4-Methylenedioxyphenyl)-2-(furan-2-yl)nitroethene.
30. A: 3-Nicotinoylindole.
    B: 3-(3-Pyridinylmethyl)indole.
31. A: 6,7-Dimethoxyisochroman-3-one.
    B: (Z)-4-(3,4-Dimethoxybenzilidene)-6,7-dimethoxyisochroman-3-one.
32. A: Benzophenone.
    B: Benzophenone oxime.
    C: Phenanthridone.
33. A: Bis-1, 3-(3,4-dimethoxyphenyl)propan-2-one.

B: 6,7-Dihydro-2,3,9,10-tetramethoxy-5H-dibenzo[a,c]cyclohepten-6-one.
34. A: Hexane-2,5-dione.
    B: 3-Methylcyclopent-2-enone.
35. A: *exo*-6-Methoxycarbonyl-7-oxabicyclo[2.2.1]heptene.
    B: *endo*-6-Methoxycarbonyl-7-oxabicyclo[2.2.1]heptene.
    C: *cis-exo*-2,3-Dihydroxy-*exo*-6-methoxycarbonyl-7-oxabicyclo[2.2.1]-heptane.
    D: *cis-exo*-2,3-Dihydroxy-*endo*-6-methoxycarbonyl-7-oxabicyclo[2.2.1]-heptane.
36. A: N-Benzyloxycarbonylthioglycylglycine methyl ester.
37. A: 2-Butyl 4-oxopentanoate.
    B: 2-Butyl 5-formylfuran-2-yl ether.
38. A: 2-Methoxycarbonyl-1-triphenylmethylaziridine.
39. A: 1,3,4,6,11,11a-Hexahydro-8,9-methylenedioxybenzo[b]quinolizin-2(2H)-one.
40. A: 4-Methylpent-4-en-2-ol.
41. A: 5-(1'-Hydroxyundecanyl)cyclopent-2-enone.
42. D: Benzyl 3-acetoxy-2-oxoazetidin-1-yl acetate.

INDEX

Ultraviolet and visible spectroscopy

Auxochromes, 1
Beer–Lambert law, 4
Bibliography, 23
Cell design, 5
Chromophores, 6
Compound types
　alkenes, 8–10
　arenes, monocyclic, 15–16
　　disubstituted, 18–19
　　polycyclic, 21–22
　carbonyl compounds, aliphatic, 10–14
　　aromatic, 19–20
　dienes, 8–10
　enones, 11–14
　α-halogenoalkyl ketones, 11
　heterocyclic compounds, 21–23
　polyenones, 11–14
Conjugation, effect on spectra, 6, 11, 16
Electron transition (ET) bands, 16
Fieser–Woodward rules
　for polyenes, 6–10
　for polyenones, 11–14
Hyperconjugation, 7
Illustrated examples
　acetanilide, 17
　acetoxy-3,4,4 4a, 6,7,8-hexahydro-4a-methylphenanthrene, 9
　4-amino-2-methylbenzoic acid, 20
　aniline, 16
　aniliniun hydrogen sulphate, 16
　anthracene, 22
　benzene, 15
　biphenyl, 21
　buta-1,3-diene, 7
　5-chloro-2-hydroxybenzaldehyde, 19
　cholesta-3,5-diene, 9
　cholesta-2,4-dien-6-one, 14
　cholesta-4,6-dien-3-one, 13
　cyclohexene, 8
　ethene, 7
　7-hydroxyindanone, 20
　indole, 23
　isoquinoline, 23
　2-methyl-4,4a,5,6,7,8-hexahydronaphthalene, 9
　1-methylnaphthalene, 22
　napththalene, 21
　4-nitrophenol, 19
　phenanthrene, 22
　phenol, 17
　phenyl acetate, 18
　pyridine, 22
　pyrrole, 22
　quinoline, 23
　sodium phenate, 18
　stilbene, 21
　3,5,5-trimethylcyclohexenone, 11
Molar adsorptivity, 5
Probability factors, 4
Scott's rules for aryl carbonyl compounds, 20
Solvent corrections, 14
Solvents, 6
Steric effects, 21

Infrared spectroscopy

Absorption of energy, 25
Absorption frequency and intensity, 27
Bibliography,
Compound types (tables and discussion)
　acyl halides, 44, 46
　alcohols, 41

248 INDEX

aldehydes, 43–44
alkanes and alkyl compounds, 30–33
alkenes, 31, 34
alkynes, 35
amides, 45, 47
amines, 42–43
amino acids, 49–50
anhydrides, 44, 46
arenes, monocyclic, 35, 37
  polycyclic, 39
azomethines, 50
carbamates, 46
carboxylic acids, 41, 47–49
carboxylates, 48
dicarboxylic acids, 41
esters, 44, 46
ethers, 51, 52
α-halogenoalkyl ketones, 44
heterocycles, 39
hydroxylated compounds, 39
isocyanates, 35, 50
isonitriles, 50
lactams, 45–46
lactones, 44, 46
ketones, 43
nitriles, 35, 50
nitro compounds, 52–53
nitroso compounds, 52–53
phenols, 41
pyridinium salts, 39
sulphur compounds, 51
water of crystallization, 41
Correlation chart, 29
Deformation bands, 33
Hooke's law, 27
Hydrogen bonding, 40–41, 43
Illustrated examples
  aniline, 42
  benzaldehyde, 44
  benzene, 35–36
  3-cresol, 40
  cyclohexanone, 43
  cyclopentanone, 44
  furan, 39
  heptanoic acid, 49
  octane, 31
  pyridine, 39
  pyrrole, 39
  thiophene, 39
Instrumentation, 24
Overtones, 37

Practical uses of IR spectra, 28
Presentation of data, 24
Problem solving, 30
Sample presentation, 25
Tautomerism, 39
Vibrational coupling, 27
Vibrational modes, 25–27

Nuclear magnetic resonance spectroscopy

Anisotropic effects, 57
Basic concepts, 85
Bibliography, 134
Bond rotation, 119–121
Chemical shift, 88–89
Chemical shifts (carbon)
  acids, 104–105
  aldehydes, 103–104
  alkanes, 98–100
  alkenes, 100–101
  alkynes, 101–102
  amides, 105–106
  arenes, 102–103
  aziridines, 119
  enones, 107
  esters, 104–105
  ketones, 103–104
  nitriles, 106
Chemical shifts (protons), 55–59, 88
  acidic protons, 127–128
  aldehydes, 124
  alkanes, 124–125
  alkenes, 126
  alkynes, 124
  arenes, 71, 74, 79, 126–127
  aziridines, 124
  cyclopropanes, 124
  oxirines, 124
  ring-systems, 131
Chiral shift reagents, 114–115
Continuous wave spectroscopy, 90
Coupling constants, 55, 60, 89–90
  C—H coupling constants, 131–132
  C—F coupling constants, 90
  H—H coupling constants, 128–131
  N—H coupling constants, 128
D.E.P.T. technique, 94–95
Deuterium exchange, 117
DMSO solvent effect, 117, 134
Dynamic effects, 115–121

Electronegativity effects, 56
Heisenberg's equation, 87
High field NMR, 107
Hybridization effects, 57
Hydrogen-bonding, 59
Illustrated examples
  aniline, 76
  aspartic acid, 68
  benzaldehyde, 77
  N-benzyl-N-methylformamide, 120–121
  (E)-2-butenal, 108
  chlorocyclohexane, 113
  chloroethane, 72
  2-chloroethanol, 70
  2-chloroethylphenyl ether, 70
  2-chloropropionic acid, 69
  4-chlorostyrene, 67
  4-chloro-2-trifluoromethylaniline, 80
  cyclohexane, 117–119
  2,3-dibromopropionic acid, 66
  1,2-dibromostyrene, 67–68
  2,6-dichlorophenol, 63
  1,4-dimethylbenzene, 75
  N,N-dimethyl-*tert*-butylamine, 119–120
  N,N-dimethylformamide, 121
  dimethyl phthalate, 78
  furan-2-carboxaldehyde, 67
  heptanol, 111
  3-iodo-4-methylaniline, 80
  methanol, 111
  1-methoxybut-1-en-3-yne, 65
  methyl 3-cyanopropionate, 70
  methyl nicotinate, 109
  4-nitroanisole, 78
  2-nitrocinnamaldehyde, 67
  2-phenylethyl acetate, 70
  phenylethylamine, 114–115
  2-phenylpropane, 73
  styrene oxide, 67
  tetramethylsilane, 132–133
  1,1,2-trichloroethane, 62
  3-(trimethylsilyl) propionic acid, 68
  vinylpiperidine, 112
Inductive effects, 56
Instrumentation, 90–93
Integral, 55
Interpretation of data
  carbon NMR, 93–97
  proton NMR, 79–85
Karplus equation, 129

Lanthanide shift reagents, 110–111
Line broadening, 87
Mesomeric effects, 74
Multiplicity of resonances, 55
N+1 rule, 59
NMR parameters, 54
Non-equivalent protons, 112–113
Nuclei in magnetic fields, 85–87
Nuclear Overhauser effects, 111–112
Nuclear spin quantum number, 92
Off-resonance decoupling, 94–97
Pascal's triangle, 59–60, 90
Problems, 121–123
Proton exchange, 115–117
Pulsed Fourier transform spectroscopy, 92–93
Quadruple effects, 109–110
Recording methods, 139
Relaxation processes, 87
Resonance effects, 58
Ring inversion processes, 117–118
Simplification of spectra, 107
Solvents, 134
Spin decoupling, 108
Spin-spin coupling, 59, 89–90
Spin states, 86–89
Splitting patterns
  A_2B, 63
  ABX, 66–68
  AMX, 64–66, 109
  AX, 60–61
  AX_2, 62–63
  AX_3, 68
  AX_6, 71
  A_2X_2, 70–71
  A_2X_3, 71
  AX_6, 71
  AA^1XX^1, 74
Tetramethyl silane as internal reference, 54–55, 132–133
Worked examples, 81–85

Mass spectrometry

Accurate mass measurement, 144–145
Analysis of ions, 137
Bar-graph, 140
Bibliography, 182
Common encountered species
  ions, 177–178
  radicals, 176
Chemical ionization, 136, 142–143

Compound types
  acetals, 164–165
  acids, 165
  aldehydes, 161–164
  alcohols, 155
  alkanes, 152–154
  alkenes, 155
  alkynes, 155
  amides, 165
  amines, 155, 159–161
  arenes, 169–173
  cyclic alcohols, 157–158
  cyclic amines, 157–158
  cyclic ketones, 164
  cyclic thiols, 157–158
  cycloalkanes, 154
  esters, 165
  ethers, 159–161
  haloalkanes, 168–169
  heteroaromatics, 174–175
  ketones, 161–165
  sulphides, 159–161
  thiols, 155
Double bond equivalent, 178
Electron ionization, 136, 142
Elimination and rearrangement
    processes, 151–152, 172–173
Even-electron ions, 149–150
Fast atom bombardment, 136, 143–144
Field desorption, 136
Fragmentation processes, 147–148, 150–151
Gas chromatography, 136, 138
Heterolytic cleavage, 150
Homolytic cleavage, 150
Illustrated examples
  bromobutane, 170
  butanamide, 166–167
  butanol, 156–157
  butyl acetate, 166–167
  2-butylamine, 156–157
  butyl ethyl ether, 160

butyl methyl sulphide, 160
2-chloropentane, 170
cyclohexane, 154
cyclohexanethiol, 158
cyclohexanone, 163
cyclohexylamine, 158
cyclohexanol, 158
dimethylethanethiol, 156–157
2,2-dimethylhexane, 153
2-ethylbutanoic acid, 166–167
1-fluorobutane, 170
2-iodobutane, 170
2-methylbutanal, 163
3-methylheptane, 153–154
N-methylisobutylamine, 160
octane, 153
pentan-2-one, 163
pyridine, 174
pyrrole, 174
Inlet systems, 135–136
Interpretation of spectra, 175–179
Isotopes
  carbon, 145, 146
  halogen, 146–147
  hydrogen, 145–146
  nitrogen, 145–146
  oxygen, 145–146
  sulphur, 146
Liquid chromatography, 136
McLafferty rearrangement, 162–163, 165–167, 172
Magnetic scanning, 137
Metastable ions, 148–149
Molecular formulae determination, 138
Molecular ion, 140–142
Odd-electron ions, 149–150
Phenonium ions, 171
Quadrupole spectrometer, 138
Resolution of spectra, 138–139
Voltage scanning, 137
Worked examples, 179–182